FLORA OF TROPICAL EAST AFRICA

AÏZOACEAE

(including Molluginaceae and Tetragoniaceae)

C. JEFFREY

Succulent or subsucculent annual or perennial herbs or subshrubs, less often shrubs. Leaves simple, alternate, opposite or verticillate, sometimes crowded, often with an expanded membranous base, exstipulate or with membranous stipules. Inflorescences cymose, loosely dichasial to umbelliform or glomerulate (flowers sometimes solitary), axillary or terminal. Flowers regular, hermaphrodite, rarely unisexual. Calyx of 5, less often 4 or 3, members, polysepalous or gamosepalous, usually herbaceous and persistent. Petals absent. Staminodes sometimes present, then often petaloid. Stamens 5–many, hypogynous or episepalous, sometimes fascicled, when definite alternate with the calyx-lobes. Ovary superior or inferior, of 2–5–many united (rarely free) carpels, or of 1 carpel; loculi as many as carpels; ovules one, few, or many per loculus; placentation basal, axile, apical or parietal but not free-central. Fruits usually capsular, loculicidal or circumscissile, sometimes indehiscent, rarely mericarpic. Seeds usually subreniform, rarely strophiolate; embryo usually curved.

A rather ill-defined family linking the *Caryophyllaceae*, *Portulacaceae* and *Phytolaccaceae* in the very natural order *Centrospermae*, and reaching its own climax in *Mesembryanthemum* and its allies.

The staminodes, when they form an obvious, well-defined and brightly-coloured series, are sometimes referred to as petals, but here, for the sake of uniformity, all these morphologically related organs are called " staminodes ", qualified where necessary by " petaloid ", as in some cases the use of the word " petal " would be most inapt; similarly the outer envelope of the flower, called the perianth by some authors, is described as the " calyx " throughout.

1. All leaves narrowly linear, either densely crowded basally and concealing the short stem, or in whorls or crowded in axillary fascicles 2
   All or at least the basal leaves not narrowly linear, nor densely crowded, but the internodes readily visible, or the leaves all basal 4
2. Leaves shortly aristate; midrib prominent; seeds one per loculus, basal; perennial herb with densely-crowded basal leaves . . . **4. Psammotropha**
   Leaves not at all aristate; midrib not prominent 3
3. Perennial herb with the leaves crowded in axillary fascicles; inflorescences simple long-pedunculate umbels; scarious stipules present . **7. Hypertelis**
   Slender annual herb with the leaves in distinct whorls; inflorescences sessile or shortly pedunculate umbels; scarious stipules absent 4

4. Stem-leaves narrowly linear and whorled, or absent and the leaves all radical; small herbs with slender rather rigid upright or ascending stems and a basal rosette of linear, oblanceolate, obovate or spathulate leaves . 6. **Mollugo**
   Stem-leaves oblanceolate to suborbicular, alternate, opposite or apparently verticillate, the leaves never all radical; habit various, usually more or less diffuse prostrate or procumbent herbs 5
5. Leaves opposite or apparently verticillate 6
   Leaves strictly alternate 11
6. Carpels free; plants, especially the leaves and calyces, close-streaked with numerous whitish short-linear raphides . . . . . . 1. **Gisekia**
   Carpels united; plants without raphides 7
7. Ovary superior; staminodes absent, or if present then forked or fimbriate at the tips, white or yellow 8
   Ovary inferior; petaloid staminodes present, obvious, lanceolate, white, pink or magenta 12. **Delosperma**
8. Calyx-segments free to the base; seeds with a distinct filiform-appendaged strophiole . 5. **Glinus**
   Calyx-segments united below into a tube; seeds estrophiolate 9
9. Style 1 . . . . 9. **Trianthema**
   Styles 2–5 10
10. Flowers glomerulate; lid of capsule splitting into 2 valves; styles always 2; seeds 4, never more . . . . 10. **Zaleya**
    Flowers solitary; lid of capsule remaining in one piece; styles 2–5; seeds usually more than 4 . . . . 8. **Sesuvium**
11. Plant viscid glandular-hairy; fruit splitting into 2 indehiscent mericarps when ripe . . . 2. **Limeum**
    Plant not viscid glandular-hairy; fruit a loculicidal capsule, or hard and indehiscent, not splitting into mericarps 12
12. Ovary superior; fruit a loculicidal capsule 13
    Ovary inferior; fruit indehiscent 14
13. Calyx-segments free to the base; plant glabrous; petaloid staminodes present, magenta . 3. **Corbichonia**
    Calyx-segments united below into a tube; plant hairy; staminodes absent . . . . 11. **Aïzoön**
14. Fruit simple; calyx-tube of hermaphrodite flowers short; herbs . . . . . 13. **Tetragonia**
    Fruit compound; calyx-tube of hermaphrodite flowers long-cylindrical; shrublets . . 14. **Tribulocarpus**

1. GISEKIA

L., Mant. 2 : 554 (1771)

Miltus Lour., Fl. Cochinch.: 302 (1790)

Diffuse procumbent or creeping annual herbs, close-streaked with numerous short-linear whitish slightly prominent raphides. Leaves opposite, sessile or shortly petiolate, linear to oblanceolate-spathulate or elliptic, entire, subsucculent. Stipules absent. Inflorescences axillary, umbelliform to laxly dichasial, sessile or pedunculate. Flowers hermaphrodite or unisexual, regular, pedicellate, small and inconspicuous. Sepals 5, free. Stamens 5–20, hypogynous, free. Carpels 3–6, or 10–15, superior, free, indehiscent, one-seeded, separating into achenes in fruit ; pericarp thin, transparent, white-papillose; styles as many as the carpels.

2 species, one African, the other extending from the Cape through tropical Africa and Asia to Malaysia.

Stamens 8–15 ; carpels 5 in hermaphrodite flowers ; sometimes flowers unisexual, then the male with 10–15 stamens, the female with 10–15 carpels . 1. *G. africana*
Stamens 5 ; carpels 5 ; all flowers hermaphrodite . 2. *G. pharnaceoïdes*

1. **G. africana** (*Lour.*) *O. Ktze.*, Rev. Gen. 3 (2) : 108 (1898) ; F.C.B. 2 : 101 (1951). Type : Mozambique Is., *Loureiro* (location of type unknown.)

A semi-succulent glabrous herb. Stems radiating from the crown, procumbent or prostrate, spreading, branching, 10–60 cm. or more in length. Leaves mostly opposite, sessile or obscurely petiolate ; blade elliptic, oblanceolate-spathulate, oblanceolate-oblong or oblanceolate, entire, 5–45 mm. long, 1–12 mm. broad, the apex retuse, obtuse, rounded or subacute, often slightly apiculate ; petiole 0–5 mm. long. Inflorescences sessile or pedunculate, umbelliform to laxly dichasial, 3–20-flowered. Pedicels 4–20 mm. long. Flowers hermaphrodite or unisexual, pinkish. Sepals 5, herbaceous, 2–3 mm. long in fruit. Stamens 10–15 in male and hermaphrodite flowers, the filaments but slightly broadened below. Carpels 5 in hermaphrodite flowers, 10–15 in female flowers.

var. **africana**

Inflorescences umbelliform, sessile or subsessile in the axils of foliage leaves.

KENYA. Lamu District : Osine, 8 Oct. 1957, *Greenway & Rawlins* 9291 ! ; Kui Is., Sep. 1955, *Rawlins* 155 !
DISTR. **K**7 ; widespread in south tropical and South Africa ; forms also occur in Ethiopia and Somaliland
HAB. Open bushland on sand-dunes, near sea, 0–25 m.

SYN. *Miltus africana* Lour., Fl. Cochinch. : 302 (1790)
Gisekia miltus Fenzl, Nov. Stirp. Dec. Vind. 10 : 86 (1839) ; F.T.A. 2 : 594 (1871) ; F.D.O.-A. 2 : 253 (1938), *nom. illegit.* Type : as *G. africana*
G. pentadecandra Moq. in DC., Prodr. 13 (2) : 28 (1849). Type : South Africa, *Zeyher* (K, iso. !)

var. **pedunculata** (*Oliv.*) *Brenan* in Mem. N. Y. Bot. Gard. 8 : 444 (1954). Type : Portuguese East Africa, Tette, *Kirk* (K, lecto. !)

Inflorescences umbelliform to laxly dichasial, pedunculate ; peduncles simple or cymosely branched, 15–50 mm. long.

TANGANYIKA. Southern Province : without locality or date, *Busse* 1052 !
ZANZIBAR. Zanzibar Is., Marahubi, 27 Dec. 1929, *Vaughan* 979 !
DISTR. **T**8 ; **Z** ; widespread in south tropical and South Africa
HAB. Maritime sands

SYN. *Gisekia aspera* Klotzsch in Peters, Reise Mossamb. Bot. : 136 (1861). Type : Portuguese East Africa, Sena River, *Peters* (B, holo.)
G. miltus Fenzl var. *pedunculata* Oliv. in F.T.A. 2 : 594 (1871)

FIG. 1. *GISEKIA PHARNACEOÏDES* var. *PHARNACEOÏDES*—**1,** habit, × 1; **2,** leaf-base, × 3; **3,** part of undersurface of leaf, showing raphides, × 16; **4,** inflorescences, × 1; **5,** bract, × 16; **6,** flower, × 6; **7,** sepal, × 10; **8,** stamen, × 16; **9,** gynoecium, × 12; **10,** fruit, × 6; **11,** achene, × 10; **12,** seed, × 10. 1–3, from *Bruce* 567; 4–12, from *Pielou* 58.

The distribution of unisexual and hermaphrodite flowers in this species needs further study.

2. **G. pharnaceoïdes** *L.*, Mant. 2 : 562 (1771) ; F.T.A. 2 : 593 (1871) ; F.W.T.A. 1 : 113 (1927) ; ed. 2, 1 : 134 (1954) ; F.D.O.-A. 2 : 253 (1938) ; Andrews, Fl. Pl. Anglo-Egypt. Sudan 1 : 91 (1950) ; F.C.B. 2 : 102 (1951). Type : from eastern India, cultivated at Uppsala (LINN, lecto.)

A semi-succulent glabrous herb, often with a pink tinge in the stems and leaves. Stems trailing, decumbent, or prostrate, 2–70 cm. or more in length. Leaves mostly opposite, sessile or obscurely petiolate ; blade linear through linear-oblanceolate and oblanceolate-spathulate to elliptic, entire, 5–60 mm. long, 1–19 mm. broad, the apex rounded, obtuse or subacute, often obscurely apiculate. Inflorescences sessile or pedunculate, umbelliform, congested to very lax, 3–40-flowered ; peduncles usually simple, sometimes cymosely branched, up to 55 mm. long. Flowers hermaphrodite, greenish or greenish-white, often with a pink, mauve or yellowish tinge. Sepals 5, 1–3 mm. long in fruiting calyx. Stamens 5 ; filaments rather broadened below. Carpels and styles 5. Fig. 1.

var. **pharnaceoïdes**

Flowers in ± congested sessile or pedunculate umbelliform cymes ; pedicels fairly stout ; calyces 1·5–3 mm. long in fruit. Fig. 1.

UGANDA. Karamoja District : Kacheliba, 20 May 1940, *A. S. Thomas* 3391 ! ; Bunyoro District : Bulisa, Jan. 1941, *Purseglove* 1094 ! ; Busoga District : Butembe Bunya, N. end of Sagitu Is., 16 Jan. 1953, *G. H. S. Wood* 642 !
KENYA. Baringo District : 19 km. N. E. of Eldama Ravine, 21 Aug. 1951, *Bogdan* 4223 ! Mombasa District : Mombasa, English Point, 1 June 1934, *Napier* 3395 *in C.M.* 6200 ! ; Kilifi District : Kibarani, 12 May 1945, *Jeffery* K190 !
TANGANYIKA. Shinyanga, *Koritschoner* 2287 ! ; Lushoto District : 8 km. SE. of Mkomazi, 2 May 1953, *Drummond & Hemsley* 2379 ! ; Singida District : near the Boma, 3 Mar. 1928, *B. D. Burtt* 1367 !
DISTR. **U**1–4 ; **K**1–4, 7 ; **T**1–6, 8 ; from South Africa and the Mascarene Is. to India and Ceylon
HAB. Ruderal and weed of cultivated ground, roadsides, grassland, bushland and woodland, especially on sandy soils, 0–1700 m.

SYN. *Gisekia linearifolia* Schumach. & Thonn., Beskr. Guin. Pl.: 167 (1827). Type : Ghana, *Thonning* (C, holo.)
G. rubella Moq. in DC. Prodr. 13(2) : 27 (1849) ; F.T.A. 2 : 594 (1871) ; Andrews, Fl. Pl. Anglo-Egypt. Sudan 1 : 92 (1950) ; F.C.B. 2 : 103 (1951). Type : Sudan, *Kotschy* 2 (K, iso. !)
G. congesta Moq. in DC., Prodr. 13(2) : 28 (1849). Type : Senegambia, *Heudelot* 458 (K, holo. !)
G. pharnaceoïdes L. var. *pedunculata* Oliv. in F.T.A. 2 : 594 (1871). Type : as *G. linearifolia* Schumach. & Thonn.
G. pharnaceoïdes L. var. *congesta* (Moq.) Oliv. in F.T.A. 2 : 594 (1871)

This species is very variable, especially in leaf-shape and degree of compactness of the inflorescence, but the variation is so continuous and reticulate that, although it is undoubtedly genetic, maintenance of any of the previously proposed varieties as distinct is impossible. However, the following seems to merit varietal status.

var. **pseudopaniculata** *Jeffrey* in K.B. 14 : 235 (1960). Type : Kenya, Northern Frontier Province, Dandu, 5 May 1952, *Gillett* 13041 (K, holo. ! ; EA, iso. !)

Flowers in laxly pedunculate few-flowered umbels : peduncles filiform, cymosely branched ; pedicels filiform, 3–14·5 mm. long : calyces 1–1·8 mm. long in fruit.

KENYA. Northern Frontier District : Wajir, 21 June 1951, *Kirrika* 57 ! ; Turkana District : 40 km. SW. of Lodwar, 12 May 1953, *Padwa* 142!
DISTR. **K**1, 2, 7 ; also in the Ogaden, Ethiopia
HAB. In *Acacia-Commiphora* bushland, on sandy soils; 550–800 m.

2. LIMEUM

L., Syst. Nat., ed. 10 : 995 (1759) ; Friedr. in Mitt. Bot. Staatss. München 14–15 : 134 (1956) ; Verdc. in K.B. 12 : 351 (1957)
Gaudinia J. Gay in Bull. Sci. Nat. 18 : 412 (1829), *non* Beauv. (1812)
Semonvillea J. Gay in Bull. Sci. Nat. 18 : 412 (1829)
Dicarpaea Presl, Symb. Bot. 1 : 37 (1830)
Acanthocarpaea Klotzsch in Peters, Reise Mossamb. Bot. : 137, t. 24 (1861)

Glabrous or viscid glandular-hairy annual or perennial herbs or subshrubs, rarely shrubs. Leaves opposite or alternate, subulate, linear, lanceolate, ovate or orbicular, entire. Stipules absent. Inflorescences cymose, terminal, often overtopped and then apparently lateral, sessile or pedunculate, few- or many-flowered, lax or glomerulate. Flowers hermaphrodite, greenish, small. Sepals 5, herbaceous, free. Staminodes 0–5, arising from the base of the outer stamens, sometimes petaloid, free. Stamens usually 7 (5 outer, 2 inner) sometimes 5, hypogynous ; filaments broadened below. Ovary superior, syncarpous, of 2 carpels, 2-locular ; style 1 ; stigmas 2. Fruit separating into 2 mericarps. Mericarps indehiscent, 1-seeded, suborbicular or reniform, often auricled at the base, the outer (rounded) face reticulate, rugose, spinescent, or smooth, wingless or sometimes broadly membranous-winged.

Some 20 species, mostly South African but with a few species extending the range through tropical Africa and Arabia to Pakistan.

Sepals without an acumen ; inflorescence 0–2·0 cm. stalked . . . . 1. *L. viscosum*
Sepals with a distinct often recurved acumen ; inflorescence 3·6–6·6 cm. stalked . . . . 2. *L. praetermissum*

1. **L. viscosum** (*J. Gay*) *Fenzl*, Nov. Stirp. Dec. 10 : 87 (1839) ; F.T.A. 2 : 595 (1871) ; F.W.T.A. 1 : 113 (1927) ; ed. 2, 1 : 134 (1954) ; F.D.O.-A. 2 : 252 (1938). Type : Senegal, *Sieber* 62 (K, iso. !)

A viscid glandular-hairy annual or short-lived perennial herb. Stems weak, diffuse, more or less divaricate, suberect, procumbent or creeping, 2·5–45 cm. long. Leaves alternate, petiolate ; blade viscid, linear-lanceolate, lanceolate, oblanceolate-spathulate, ovate or almost orbicular, 5–35 mm. long, 1·5–17 mm. broad, entire, retuse, rounded or obtuse, sometimes obscurely mucronulate at apex, more or less cuneate at base ; petiole 1–10 mm. long. Inflorescences apparently lateral, extra-axillary, glomerulate or fairly lax, pedunculate or sessile, 5–25-flowered. Peduncles 0–2·0 cm. long. Sepals 5, subherbaceous, unequal, broadly ovate. Staminodes 0–5, evanescent, clawed with expanded limb, the limb broadly ovate, cuneate-truncate at base, equalling sepals or much shorter. Stamens 5 or 7, inserted on hypogynous disc. Ovary 2-lobed. Mericarp 2–5 mm. diameter, 1-celled, 2-auricled, (auricles smooth), wingless, marked on outer (rounded) face with a network of ridges forming a honeycomb pattern, more or less prominently elevated into small projections at the angles ; this pattern sometimes almost obsolete. Fig. 2.

subsp. **viscosum**

Staminodes 0, or if present then much shorter than sepals ; inflorescence pedunculate or if sessile then leaves lanceolate to ovate, not broadly ovate to orbicular. Fig. 2/11–13.

var. **viscosum**

Peduncle 1–2 cm. long ; stamens 7 ; leaves up to 2·5 times as long as broad ; pedicels 2–3·5 mm. long ; mericarp 2–3 mm. dia. Fig. 2/11–13.

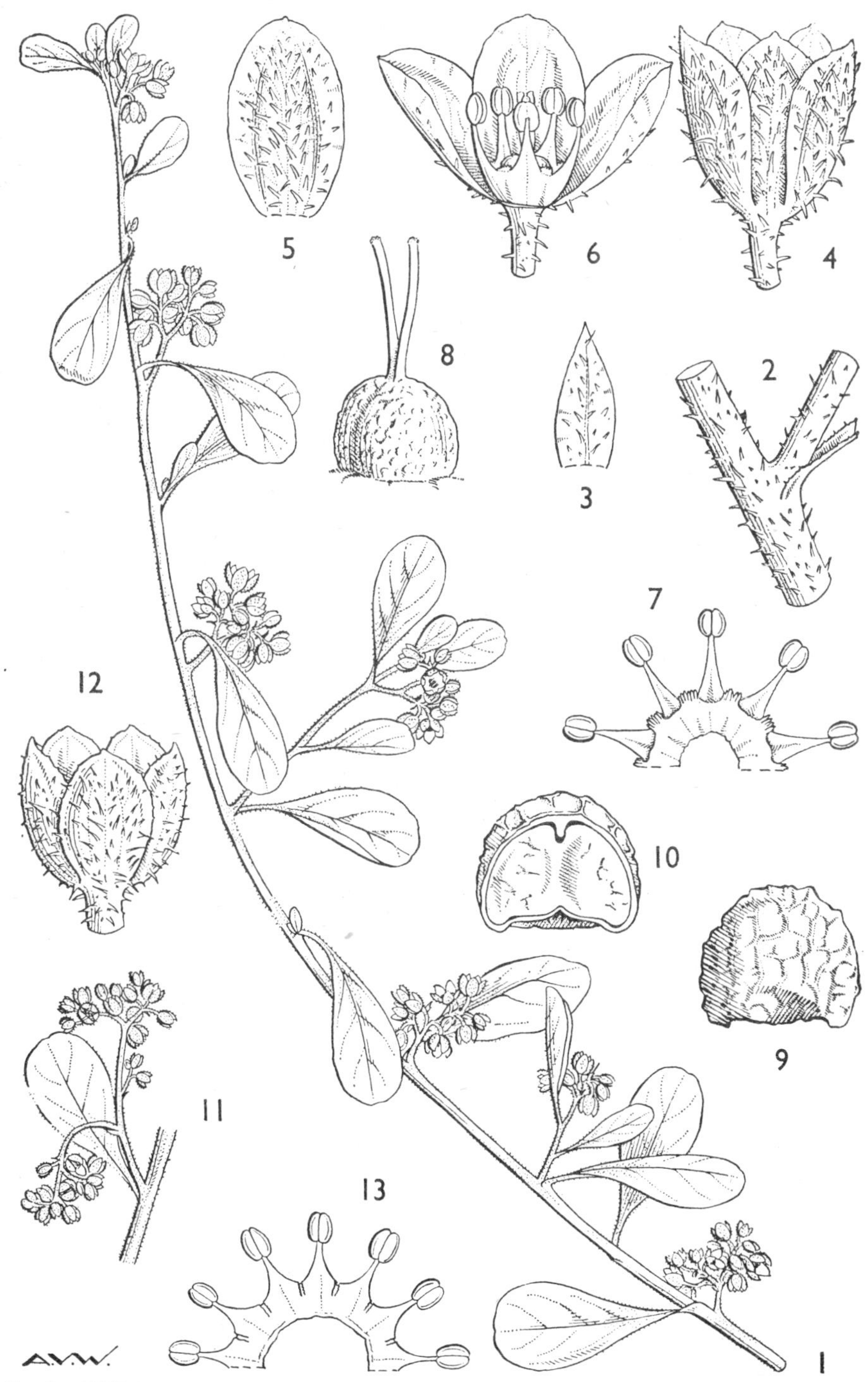

FIG. 2. *LIMEUM VISCOSUM* subsp. *VISCOSUM* var. *KENYENSE*—**1,** habit, × 1; **2,** leaf-base, × 4; **3,** bract, × 8 ; **4,** flower, × 8 ; **5,** sepal, × 8 ; **6,** flower with 2 sepals removed, showing stamens, × 8; **7,** androecium, × 8 ; **8,** ovary, × 20 ; **9,** mericarp, outer face, × 10 ; **10,** mericarp, commissural face, × 10 ; var. *VISCOSUM*—**11,** inflorescence, × 1 ; **12,** flower, × 8 ; **13,** androecium, × 10. 1–10, from *Napier* 1000 ; 11-13, from *B. D. Burtt* 1292.

TANGANYIKA. Kondoa District : near Kolo, 1 Feb. 1928, *B. D. Burtt* 1292 ! ; Mpwapwa District : Kongwa, 10 Mar. 1950, *Anderson* 645 ! ; Kilosa District : Kibedya near Kilosa, Jan. 1931, *Haarer* 1945 !
DISTR. **T**3, 5, 6 ; Angola, Southern Rhodesia ; Ethiopia, Sudan, N. Nigeria, Senegal
HAB. Weed of cultivated places; 750–1440 m.

SYN. *Gaudinia viscosa* J. Gay in Bull. Sci. Nat. 18 : 412 (1829)
Limeum viscosum (J. Gay) Fenzl var. *kotschyi* Moq. in DC., Prodr. 13(2) : 23 (1849). Type : Sudan, *Kotschy* 20 (K, iso. !)
L. kotschyi (Moq.) Schellenb. in E.J. 48 : 497 (1912) ; Andrews, Fl. Pl. Anglo-Egypt. Sudan 1 : 94 (1950)

var. **kraussii** *Friedr.* in Mitt. Bot. Staatss. München 14–15 : 152 (1956). Type : Natal, *Krauss* 135 (M, holo.)

Peduncle 1–2 cm. long ; stamens 7 ; leaves 3–5 times as long as broad.

TANGANYIKA. Lindi District : Lake Lutamba, *Schlieben* 5963 !
DISTR. **T**8 ; Portuguese East Africa, Natal, Transvaal
HAB. Open woodland ; 250 m.

SYN. *Limeum natalense* Schellenb. in E.J. 48 : 495 (1912). Type : Natal, *Kuntze* (B, holo.)

var. **kenyense** *Friedr.* in Mitt. Bot. Staatss. München 14–15 : 153 (1956). Type : Kenya, Teita District, Ndara, *Hildebrandt* 2400 (B, holo., K, iso. !)

Peduncle up to 0·5 cm. long ; stamens usually 5. Fig. 2/1–10, p. 7.

UGANDA. Karamoja District : Pian, 21 July 1958, *Mrs. Dyson-Hudson* 465 !
KENYA. Machakos District : Makindu River, 14 Apr. 1902, *Kassner* 581 ! ; Kiambere, 24 Nov. 1951, *Kirrika* 140 ! Teita District : Voi, 9 May 1951, *Napier* 1000 !
TANGANYIKA. Lushoto District : 8 km. SE. of Mkomazi, 2 May 1953, *Drummond & Hemsley* 2383 !
DISTR. **U**1 ; **K**2, 4, 7 ; **T**3 ; not known elsewhere
HAB. Open bushland, river-beds and a weed of cultivated places ; 450–750 m.

SYN. *Limeum orientale* Schellenb. in E.J. 48 : 497 (1912). Type : as var. *kenyense*

2. **L. praetermissum** *Jeffrey* in K.B. 14 : 235 (1960). Type : Kenya, Northern Frontier Province, Wajir, *Gillett* 13371 (K, holo. !)

A viscid glandular-hairy perennial herb. Stem more or less prostrate, becoming when older thickened and subwoody with a smooth white bark, 40 cm. long. Leaves alternate, petiolate. Blade viscid, obovate, elliptic or oblanceolate, 15–26 mm. long, 7–12 mm. broad, entire, retuse or rounded at the apex, narrowed into the petiole below. Petiole 2–4 mm. long, somewhat membranous-winged towards the base. Inflorescences leaf-opposed, laxly branched, 15–35-flowered, on peduncles 3·6–6·6 cm. long. Pedicels 2–6 mm. long. Bracts and bracteoles small, linear-lanceolate, acuminate, membranously margined. Flowers greenish-white. Sepals 5, herbaceous, green with white membranous margins, ovate, 5–6 mm. long, with a prominent often recurved acumen. Staminodes 0. Stamens 5, free, filaments broadly dilated below. Ovary obscurely 2-lobed, glabrous. Mericarps about 3 mm. in diameter, wingless, smooth on the outer (upper) rounded surface except for a series of short grooves and ridges submarginal and perpendicular to the commissural faces, at the base obscurely 2-auricled laterally, auricles smooth.

KENYA. Northern Frontier Province : Wajir, 27 May 1952, *Gillett* 13371 !
DISTR. **K**1 ; not known elsewhere
HAB. Open places in overgrazed *Commiphora* scrub, 240 m.

3. CORBICHONIA

Scop., Introd. : 264 (1777)
Orygia Forsk., Fl. Aegypt.-Arab. : 103 (1775), pro parte, excl. typ.

Glabrous procumbent or ascending annual or perennial herbs. Leaves alternate, petiolate, obovate to rotundate, entire, apiculate, subsucculent. Inflorescences terminal or leaf-opposed few- to many-flowered cymes. Flowes hermaphrodite, regular, pedicellate. Sepals 5, free. Staminodes many, delicate, petaloid. Stamens many, hypogynous. Ovary superior, 5-locular, 5-lobed, of 5 united carpels ; ovules many per loculus, axile ; styles 5, free, sessile, linear. Fruit a 5-valved loculicidal capsule. Seeds many, subreniform.

2 species, one in tropical Africa and Asia, the other confined to South West Africa.

C. decumbens (*Forsk.*) *Exell* in J.B. 73 : 80 (1935) ; Andrews, Fl. Pl. Anglo-Egypt. Sudan 1 : 91, fig. 59 (1950) ; F.C.B. 2 : 104 (1951). Type : Arabia, *Forskål* (C, holo., BM, iso. !)

An erect, decumbent or prostrate glabrous annual or short-lived perennial herb up to 45 cm. high. Stems branched, wiry, often (like the rootstock) subwoody at the base, up to 100 cm. long, the internodes with narrow ridges decurrent from the petiole-bases. Leaves petiolate ; blade oblanceolate-obovate, obovate-spathulate, or obovate, 4·5–65 mm. long, 2–33 mm. broad, entire, subsucculent, glaucous, apex apiculate, base cuneate ; the petiole winged, 1·5–11 mm. long. Inflorescences terminal but overtopped and thus appearing lateral, leaf-opposed, few- to many-flowered, lax, racemiform, 1–17 cm. long. Bracts lanceolate, membranous, up to 3 mm. long. Pedicels 2–3 mm. long. Flowers pink, mauve, magenta or red. Sepals herbaceous, green with white membranous margins, broadly ovate, about 4 mm. long in fruit. Staminodes many, petaloid, delicately membranous, fugaceous, at first shorter, then longer than the sepals. Stamens many, in 2 rings. Fruit yellow-green, shining. Fig. 3, p. 10.

UGANDA. West Nile District : Nebbi, Sept. 1940, *Purseglove* 1061 ! ; Toro District : Semliki Plains, 23 Nov. 1955, *A. S. Thomas* 1542 ! ; Teso District : Serere, Dec. 1931, *Chandler* 300 !

KENYA. Turkana District : Lodwar flats, Feb. 1940, *Leakey in Bally* 1134 ! ; Teita District : Voi, 8 May 1931, *Napier* 989 ! & Tsavo Station, 13 Feb. 1953, *Sheldrake in Bally* 8567 !

TANGANYIKA. Lushoto District : Mazinde, 1 Nov. 1953, *Drummond & Hemsley* 2332 ! Uzaramo District : Dar es Salaam, *Marshall* 15 ! ; Masasi District : between Matambe and Marumba on Ruvuma R., 23 Feb. 1901, *Busse* 1382 !

DISTR. **U**1, 3–4 ; **K**1–5, 7 ; **T**2–4, 6, 8 ; widespread in drier parts of tropical Asia and Africa ; introduced into tropical America

HAB. Open bushland, grassland and also a weed of roadsides and of cultivation, especially on sandy soils ; 0–1080 m.

SYN. *Orygia decumbens* Forsk., Fl. Aegypt.-Arab.: 103 (1775) ; F.T.A. 2 : 589 (1871) ; F.D.O.-A. 2 : 255 (1938)

Glinus trianthemoïdes Heyne in Roth, Nov. Pl. Sp.: 231 (1821). Type : E. India, *Heyne* (B, holo.)

Axonotechium trianthemoïdes (Heyne) Fenzl in Ann. Wien. Mus. 1 : 355 (1836)

Orygia mucronata Klotzsch in Peters, Reise Mossamb. Bot.: 140 (1861). Type : Portuguese East Africa, Tette, *Peters* (B, holo.)

Glinus mucronatus (Klotzsch) Klotzsch in Peters, Reise Mossamb. Bot.: 570, t. 25. (1864)

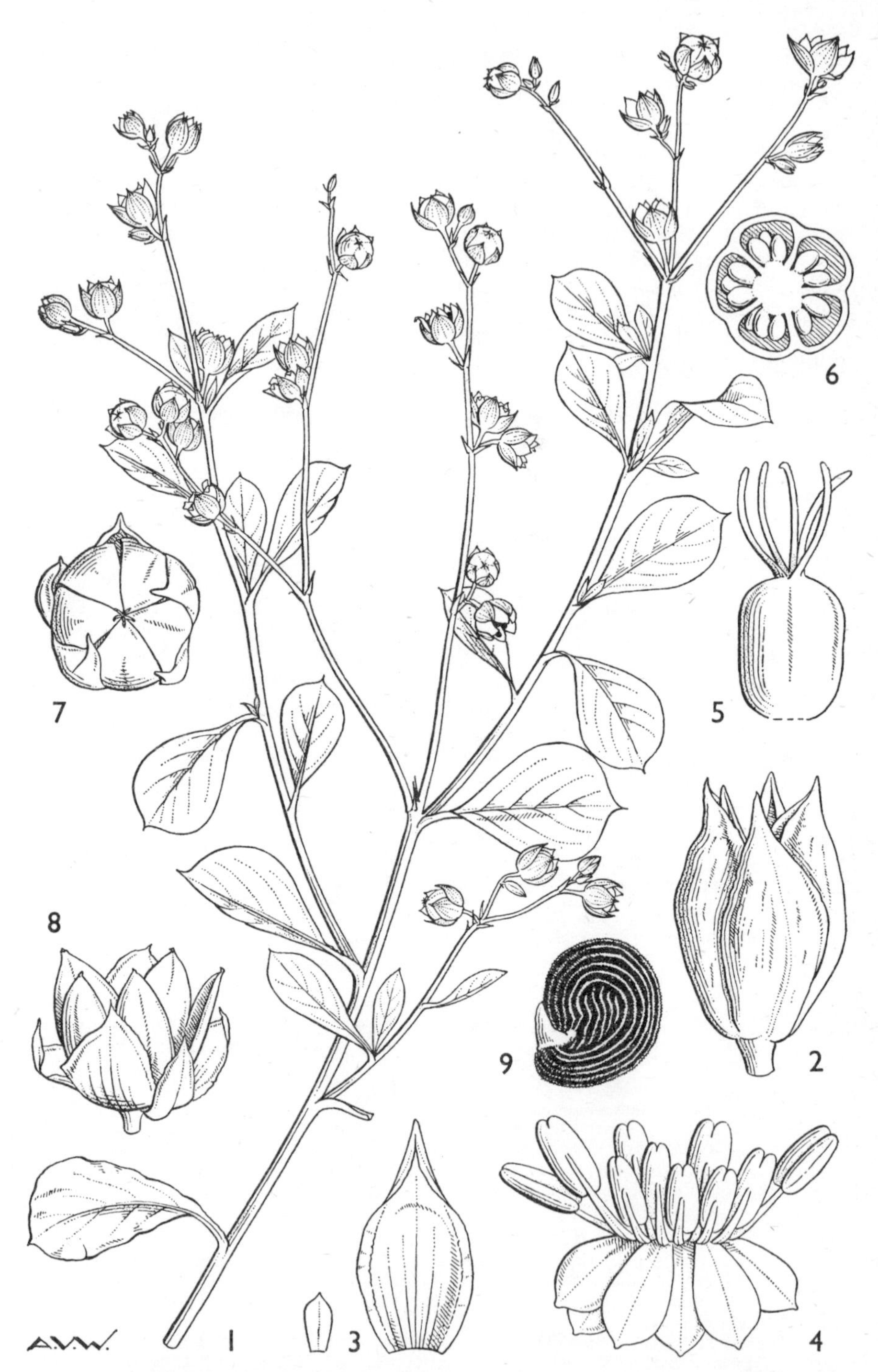

FIG. 3. *CORBICHONIA DECUMBENS*—**1**, habit, × 1 ; **2**, flower, × 8 ; **3**, sepal and young staminode, × 8 ; **4**, part of rings of staminodes and stamens, × 16 ; **5**, ovary, × 12 ; **6**, ovary, in transverse section, × 16 ; **7**, capsule, in plan, × 4 ; **8**, fruiting calyx, × 4 ; **9**, seed, × 20. 1, from *Bogdan* 4321 ; 2, 5, 6, from *Martin* ; 3–4, from *Marshall* 15 ; 7–9, from *Purseglove* 1061.

4. PSAMMOTROPHA

Eckl. & Zeyh., Enum.: 286 (1836) ; Fenzl in Ann. Wien. Mus. 2 : 263 (1839) ; Adamson in J. S. A. Bot. 25 : 51 (1959)

Perennial often creeping or rosette herbs or small branched shrublets. Leaves alternate, opposite or verticillate, often densely crowded basally, at the nodes, or all along the stem, oblanceolate, linear, subulate or ericoid. Stipules scarious when present. Inflorescences lax, or umbelliform and often glomerulate terminal or axillary cymes. Flowers hermaphrodite, pedicellate, small, greenish. Sepals 5, herbaceous, free. Stamens 5, hypogynous, free, alternate with sepals. Ovary of 3–5 united carpels, 3–5-lobed, 3–5-locular ; stigmas 3–5, free or united ; ovules 1 per loculus, basal. Fruit a 3–5-lobed, 3–5-celled, loculicidal capsule. Seeds granular.

11 species, most in the eastern parts of South Africa, but 2 species extending into tropical Africa.

P. myriantha *Sond.* in Harv. & Sond., Fl. Cap. 1 : 147 (1860) ; F.C.B. 2 : 112 (1951). Type : South Africa, Transvaal, *Zeyher* 616 (S, holo. ; BM, K, iso. !)

A low tufted herb, with densely-crowded linear basal leaves hiding the stem, sometimes offsetting by means of runners rooting at the nodes. Stems thick, short, erect, persistent, mostly subterranean, branched or unbranched, 1–8 cm. long. Basal leaves sessile, with broad membranous bases, linear, entire, strongly 1-nerved, glabrous, often with recurved margin, shortly aristate, 4–90 mm. long. Flowering stems annual, similar in appearance to the runners, ascending, glabrous, solitary or several, 0·5–2·5 mm. diameter at the base, with internodes up to 1·5 cm. long, leafy, simple or much branched in a divaricate cymose manner, bearing whorls of leaves and flowers at the nodes. Stem-leaves stipulate, sessile, linear or lanceolate, entire and rather glaucous, shortly aristate, 2–3·5 mm. long. Stipules persistent, linear. Flowers creamy or greenish, pedicellate, in sessile umbelliform cymes clustered at the nodes. Pedicels 2–5 mm. long. Sepals 5, broadly ovate, green, with wide membranous margins, 2–2·5 mm. long in flower. Ovary 5-(rarely 3-) locular, of 5 (rarely 3) carpels ; fruit a 5- (rarely 3-) lobed capsule, depressed above, splitting into as many triangular valves. Fig. 4, p. 12.

TANGANYIKA. Njombe District : Elton Plateau, May 1953, *Eggeling* 6618 ! ; Kipengere Mts., 7 Jan. 1957, *Mrs. Richards* 7585 !
DISTR. **T7** ; SE. Belgian Congo, Northern & Southern Rhodesia, Angola and South Africa
HAB. Crevices in rock-outcrops ; 2400–2600 m.

SYN. *Psammotropha myriantha* var. *huillensis* Oliv. in F.T.A. 2 : 593 (1871). Type : Angola, *Welwitsch* 2417 (K, holo. !)
P. breviscapa Burtt-Davy, Man. Fl. Pl. Transv. 1 : 49 (1926). Type : South Africa, Transvaal, *M. Wood* 5728 (K, holo. !)

5. GLINUS

L., Sp. Pl. : 463 (1753) & Gen. Pl., ed. 5 : 208 (1754)

Paulo-Wilhelmia Hochst. in Flora 27 : 17 (Jan. 1844), *non* Hochst. in Flora 27, Beil. : 4 (1844, after Jan.)

Often prostrate simply or stellately pubescent or almost glabrous somewhat succulent annual herbs. Leaves opposite or apparently verticillate,

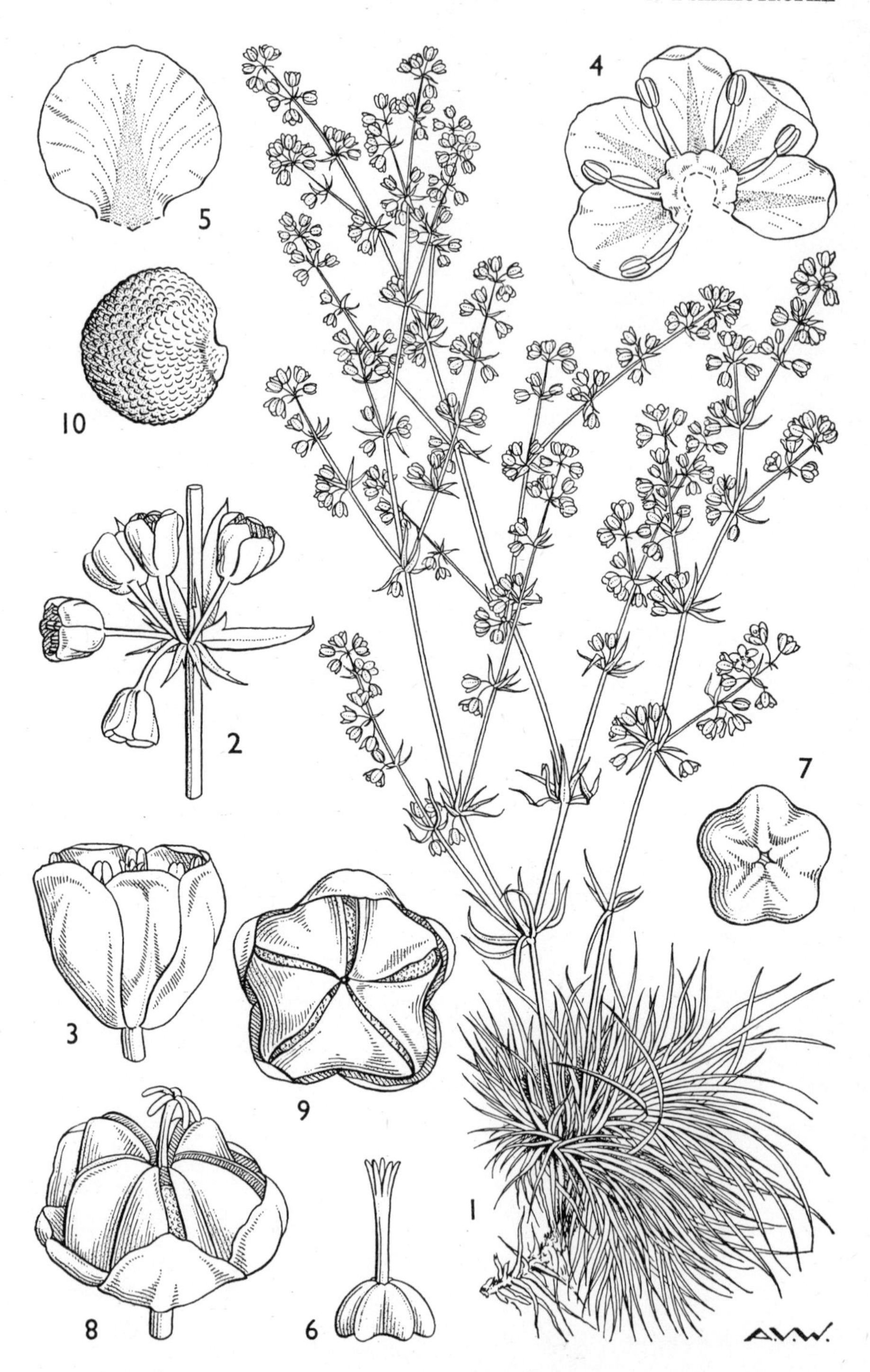

FIG. 4. *PSAMMOTROPHA MYRIANTHA*—**1**, habit and inflorescences, × 1 ; **2**, part of inflorescence, × 4 ; **3**, flower, × 12 ; **4**, androecium, × 8 ; **5**, sepal, × 12 ; **6**, ovary, × 12 ; **7**, ovary, in plan, × 20 ; **8**, fruiting calyx, × 12 ; **9**, capsule, in plan, × 12 ; **10**, seed, × 24. 1, 8–10, from *Eggeling* 6618 ; 2–7, from *Paulo* 291.

narrowly elliptic to almost orbicular, entire or serrate, shortly petiolate. Flowers fascicled at the nodes, axillary, pedunculate. Sepals 5, herbaceous, free. Staminodes 0–20 or more, usually divided at the apex, sometimes petaloid. Stamens 3–30 or more, free, hypogynous. Ovary superior, of 3–5 united carpels, 3–5-locular, the loculi multiovulate ; placentation axile; stigmas sessile, free, as many as carpels. Fruit a 3–5-valved loculicidal capsule. Seeds granulate or smooth, each with a distinct long filiform-appendaged strophiole.

About 6 species, 4 of which are African; found throughout the tropics and subtropics.

Plant very nearly glabrous, glabrescent or sparsely-hairy, especially in younger parts ; hairs stellate or simple :
  Hairs simple, crisped :
    Fruiting calyx shorter than or at most equalling the pedicel :
      Mature stem-leaves 3–7 times as long as broad (incl. petiole) ; pedicels 4·5–18·5 mm. long . 1. *G. oppositifolius*
      Mature stem-leaves 2–2·8 times as long as broad (incl. petiole) ; pedicels 3–5 mm. long . 2 × 1. *G. lotoïdes* × *G. oppositifolius*
    Fruiting calyx 1·3–1·5 times as long as pedicel . 2 × 1. *G. lotoïdes* × *G. oppositifolius*
  Hairs stellate ; arms of hairs straight or crisped . 2 × 1. *G. lotoïdes* × *G. oppositifolius*
Plant densely stellate-pubescent, appearing more or less felted and greyish in colour :
  Fruiting calyx 2·5–4·5 × 5–8 mm. ; staminodes simply bifid, or 0 ; flowers greenish-white . 2. *G. lotoïdes*
  Fruiting calyx 6·5–8·5 × 7–13 mm. ; staminodes multifid ; flowers yellow . . . . . 3. *G. setiflorus*

1. **G. oppositifolius** (*L.*) *A. DC.* in Bull. Herb. Boiss., sér. 2, 1 : 559 (1901) ; F.W.T.A. 1 : 114 (1927) ; ed. 2, 1 : 135 (1954) ; Andrews, Fl. Pl. Anglo-Egypt. Sudan 1 : 93, fig. 60 (1950) ; F.C.B. 2 : 118 (1951). Type : Ceylon, *Hermann* (BM, holo. !)

An upright, spreading or prostrate very nearly glabrous diffuse subsucculent herb ; branches 5–80 cm. long. Leaves opposite or apparently verticillate, petiolate, 10–51 mm. long (including petiole of 0–5 mm.), 4–12·5 mm. broad, the blade narrowly elliptic, oblanceolate-elliptic, oblanceolate-oblong or oblong-ovate, plane, entire or denticulate, subacute or acute and obscurely apiculate at apex. Flowers inconspicuous, greenish or pinkish white, in rather lax fascicles at the nodes, 1–15 per node, the pedicels 4·5–18·5 mm. long. Sepals 5, free ; staminodes 0–4, bifid ; stamens 4–10. Ovary of 3 united carpels. Fruiting calyx 1·5–2·5 mm. broad, 3·5–5 mm. long. Fig. 5/8–9, p. 14.

KENYA. Machakos District : Kiambere, 20 Nov. 1951, *Kirrika* 154 ! ; Kilifi, 13 Feb. 1946, *Jeffery* K466 !

TANGANYIKA. Lushoto District : Mkomazi, about 3 km. NE. of Lake Manka, 1 May 1953, *Drummond & Hemsley* 2347 ! ; Uzaramo District : Dar es Salaam, Nov. 1892, *Holst* 4149 ! ; Rufiji District : Utete by R. Rufiji, 2 Dec. 1955, *Milne-Redhead & Taylor* 7527 !

ZANZIBAR. Zanzibar Is., July 1873, *Hildebrandt* 907 ! ; Pemba, near Ole, 26 Oct. 1929, *Vaughan* 899 !

DISTR. **K**4, 7 ; **T**2–4, 6, 8 ; **Z** ; **P** ; pantropical

HAB. Dry river-beds and -banks and by open water, especially on sandy soils ; also a weed of cultivation ; 0–1000 m.

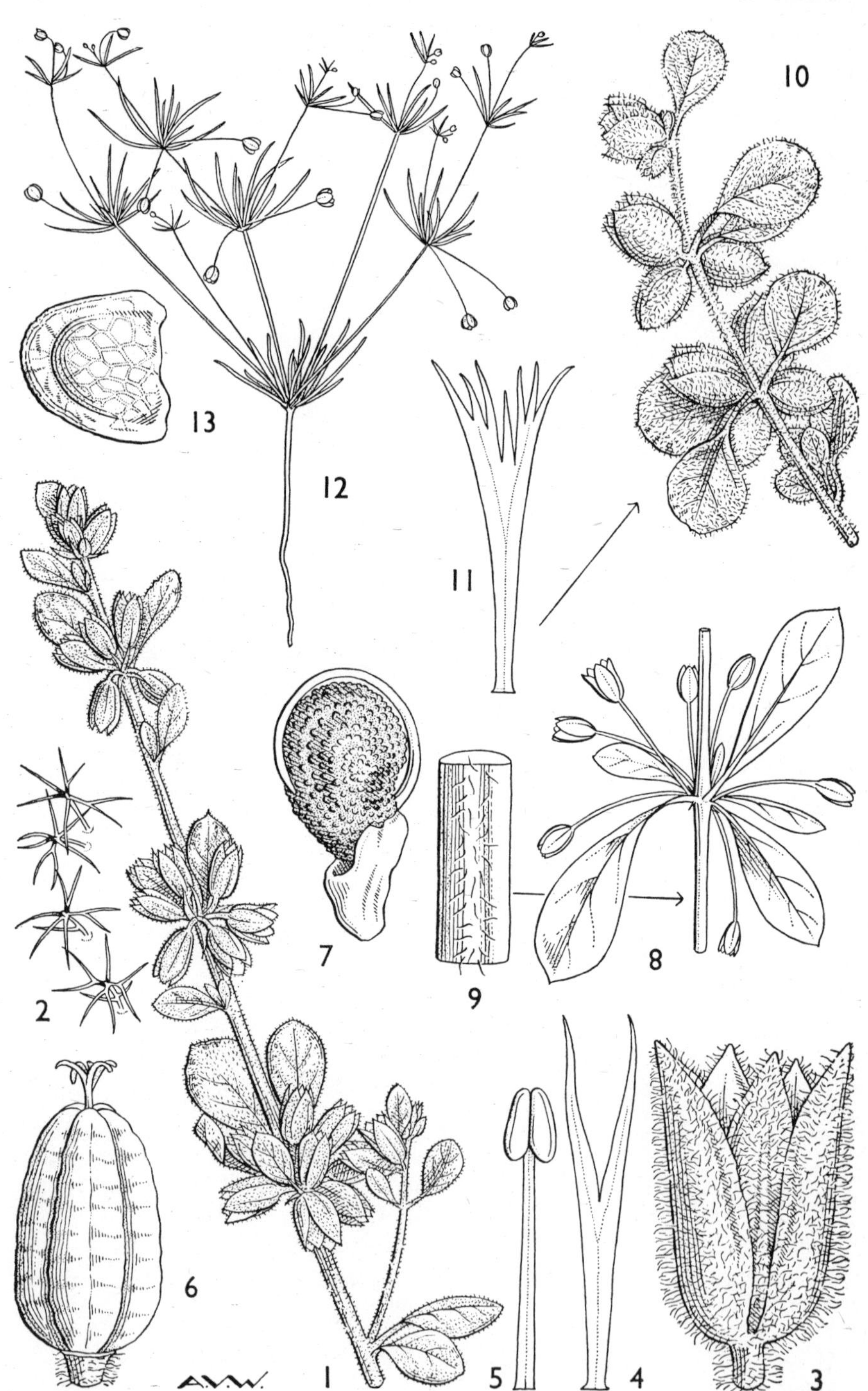

FIG. 5. *GLINUS LOTOÏDES*—**1,** habit, × 1; **2,** hairs from leaf, × 20; **3,** flower, × 6; **4,** staminode, × 10; **5,** stamen, × 10; **6,** ovary, × 8; **7,** seed, × 40; *G. OPPOSITIFOLIUS*—**8,** inflorescence, × 1; **9,** part of stem showing indumentum, × 4; *G. SETIFLORUS*—**10,** habit, × 1; **11,** staminode, × 8; *MOLLUGO CERVIANA* var. *CERVIANA*—**12,** habit, × 1; **13,** seed, × 60. 1–2, from *B. D. Burtt* 1869; 3–7, from *Geilinger* 2968; 8–9, from *Verdcourt* 1886; 10–11, from *Bally* 4506; 12, from *Gillett* 12986; 13, from *Rounce* 75.

SYN. *Mollugo oppositifolia* L., Sp. Pl.: 89 (1753)
M. spergula L., Syst., ed. 10 : 881 (1759); F.T.A. 2 : 590 (1871). Type : India, (LINN, lecto.)
Pharnaceum mollugo L., Mant. 2 : 561 (1771), *nom. illegit.* Type : as *M. spergula*
Mollugo denticulata Guill. & Perr., Fl. Seneg. Tent. 1 : 45 (1831). Type : Senegal, *Perrottet* (P, holo., BM, iso. !)
Glinus mollugo (L.) Fenzl in Ann. Wien. Mus. 1 : 359 (1836)
G. denticulatus (Guill. & Perr.) Fenzl in Ann. Wien. Mus. 1 : 361 (1836)
G. spergula (L.) Steud., Nom., ed. 2, 1 : 688 (1840) ; F.D.O.-A. 2 : 257 (1938)
Mollugo glinoïdes A. Rich., Tent. Fl. Abyss. 1 : 48 (1847). Type : Ethiopia, Shire, *Quartin-Dillon* (P, holo.)
M. serrulata Sond. in Linnaea 23 : 15 (1850). Type : South Africa, Natal, *Gueinzius* 138 (S, holo.)

2. **G. lotoïdes** *L.*, Sp. Pl. : 463 (1753) ; F.W.T.A. 1 : 114 (1927) ; ed. 2, 1 : 135 (1954) ; F.D.O.-A. 2 : 256 (1938) ; Andrews, Fl. Pl. Anglo-Egypt. Sudan 1 : 92 (1950) ; F.C.B. 2 : 107 (1951). Type : Sicily, *Boccone* (OXF, typolecto.)

A semi-erect, decumbent or prostrate and rosette-forming pale- or grey-green diffusely-branched densely stellate pubescent herb up to 20 cm. high ; branchlets 10–45 cm. long. Leaves opposite or apparently verticillate, petiolate, 10–42 mm. long (including petiole of 1–11 mm.), 5–22 mm. broad ; blade elliptic or obovate or suborbicular, plane or the veins impressed above and prominent below, entire or the margin obscurely wavy, subacute to rounded at the apex, cuneate at the base. Flowers greenish-white, sometimes with a pink tinge, not conspicuous, in tight nodal fascicles, 2–10 per node, pedicellate ; pedicels 1·5–4 mm. long. Sepals 5, free. Staminodes 0–9, strap-shaped, two-pronged at the apex. Stamens 11–30. Ovary of 5, rarely 3 united carpels. Fruiting calyx 2·5–4·5 mm. broad, 5–8 mm. long. Fig. 5/1–7.

UGANDA. Karamoja District : Kakamongole, at base of Mt. Debasien, Jan. 1936, *Eggeling* 2067 !
KENYA. Northern Frontier Province : about 24·5 km. from Laisamis on road to Marsabit, 3 Oct. 1957, *Bally* 5471 ! ; Masai District : about 1·5 km. S. of Kajiado, along Namanga road, 21 Feb. 1953, *Drummond & Hemsley* 1249 !
TANGANYIKA. Ufipa District : R. Kamba near Kampunda, 16 Oct. 1950, *Bullock* 3430 ! ; Kondoa District : Bubu Valley, on road to Salia, 8 Jan. 1928, *B. D. Burtt* 1869 ! ; Morogoro District : Ngerengere R., Nov. 1949, *Semsei in Wigg* !
DISTR. **U**1 ; **K**1–4, 6, 7 ; **T**1–8 ; widespread in the tropics and subtropics
HAB. Dry river-beds and by open water in scrub and bushland, especially on sandy soils ; also a weed of cultivation ; 0–1650 m.

SYN. *Glinus dictamnoïdes* Burm. f., Fl. Ind. : 113 (1768). Type : India, (G, holo.)
Mollugo glinus A. Rich., Tent. Fl. Abyss. 1 : 48 (1847) ; F.T.A. 2 : 590 (1871). Types : Ethiopia, R. Takazze, *Quartin-Dillon* and *Schimper* (P, syn.)

2 × 1. **G. lotoïdes** *L.* × **G. oppositifolius** (*L.*) *A. DC.* and derivatives

Hybrids of the above parentage often cause trouble in determination, although the typical forms of each species are easily distinguished. Difficulty is made greater because the hybrids are inter-fertile with each parent, so that introgression of hereditary material from each one into the other produces a wide and variable range of hybrid biotypes ; and because the hybrids can be self-fertilized and thus biotypes with distinct appearance can persist through more than one generation. Because of the introgression, although typical specimens of the species are completely separable by the above key, it is evident that the boundaries between the categories " *lotoïdes* ", " *oppositifolius* " and " *hybrid* " cannot be other than arbitrary ; and they are here so drawn that plants distinct in appearance from the typical forms of the two species can be confidently determined as hybrids or hybrid derivatives ; any observations on populations of such plants will be welcome.

G. lotoïdes L. var. *denudatus* Peter, F.D.O.-A. 2, Descr. : 28 (1932) ; F.D.O.-A. **2** : 257 (1938) (type : Tanganyika, Dodoma District, Ugogo, W. of Dodoma, *Peter* 45688 (B, holo. †)) is probably one of the hybrid forms.

The following species is quite distinct and there is no evidence as yet of its hybridization with either of the foregoing species.

3. **G. setiflorus** *Forsk.*, Fl. Aegypt.-Arab. : 95 (1775). Type : Arabia, *Forskål* (C, holo., BM, iso. !)

A suberect procumbent or prostrate grey-green densely stellate-tomentose ? annual herb up to 30 cm. high ; branches up to 60 cm. long. Leaves opposite or apparently verticillate, petiolate, 7–34 mm. long (including petiole 1–10 mm.) 6–24 mm. broad ; blade broadly obovate or suborbicular, with the veins prominent below, entire or slightly wavy at the margin, rounded or occasionally subapiculate at apex. Flowers axillary, 1–4 per node, yellow, subsessile. Sepals 5, the outer very broadly ovate and plane, the inner conduplicate. Staminodes 9–18 (in flowers dissected) yellow, petaloid, fimbriate at the apex ; stamens many (12–30 in flowers dissected). Carpels 5. Seeds granular, with a long filiform strophiolar appendage. Fig. 5/10–11, p. 14.

KENYA. Northern Frontier Province : Banessa, 22 May 1952, *Gillett* 13257 ! ; about 14 km. from Garba Tula, 18 July 1952, *Bally* 8224 ! ; Tana River District : south of Garissa, 26 Sept. 1957, *Greenway* 9236 !

TANGANYIKA. Dodoma District : Mperembwa, about 46 km. S. of Dodoma, 15 Aug. 1928, *Greenway* 773 !

DISTR. **K**1, 7 ; **T**5 ; Ethiopia, Somaliland, Arabia

HAB. Dry river-beds and near standing water in open scrub, bushland and woodland, 1000 m.

SYN. *Glinus lotoïdes* L. var. *setiflorus* (Forsk.) Fenzl in Ann. Wien. Mus. 1 : 358 (1836)
Mollugo setiflora (Forsk.) Chiov. in Bull. Soc. Bot. Ital. 1923 : 114 (1923)

6. MOLLUGO

L., Sp. Pl. : 49 (1753) & Gen. Pl., ed. 5 : 39 (1754) ; Fenzl in Ann. Wien. Mus. 1 : 375 (1836) ; Adamson in J. S. Afr. Bot. 24 : 13 (1958)

Erect or procumbent annual herbs. Leaves verticillate or sometimes opposite, linear, oblanceolate, obovate or spathulate, glabrous, entire, the radical often forming a basal rosette, the cauline sometimes lacking. Stipules small or absent, often caducous. Inflorescences cymose, laxly dichasial or umbelliform. Flowers hermaphrodite, greenish, inconspicuous. Sepals 5, free. Stamens 3–5, rarely 6–10, hypogynous, free. Ovary superior, of 3–5 united carpels, 3–5-locular ; ovules axile, many per loculus. Fruit a loculicidal capsule. Seeds many, estrophiolate.

About a dozen species; throughout the tropics and subtropics and also in warmer temperate parts; 7 species in Africa and Madagascar.

Plant both with basal leaves and with whorls of linear stem-leaves, stem-leaves persistent ; basal leaves often withered away by flowering time . . 1. *M. cerviana*
Plant with basal leaves only, the stems bearing merely minute bracts ; basal leaves persistent . 2. *M. nudicaulis*

1. **M. cerviana** (*L.*) *Ser.* in DC., Prodr. 1 : 392 (1824) ; F.T.A. 2 : 591 (1871) ; F.W.T.A. 1 : 114 (1927) ; ed. 2, 1 : 135 (1954) ; F.D.O.-A. 2 : 259 (1938) ; Andrews, Fl. Pl. Anglo-Egypt. Sudan 1 : 94 (1950) ; F.C.B. 2 : 111 (1951). Type : Rostov on Don, U.S.S.R., *Gerber* (LINN, lecto.)

A small glabrous herb with many slender rather rigid pale brownish upright or ascending stems 4–17 cm. long. Leaves sessile, often glaucous ; the basal rosetted, linear, oblanceolate, or spathulate, often quickly withering, 2–26 mm. long, 0·3–3 mm. wide ; the cauline whorled, linear, up to 18 mm. long and 1·2 mm. wide. Inflorescences axillary or terminal, sessile or pedunculate, umbelliform, 1–4-flowered. Peduncle up to 20 mm. long. Pedicels 5–15 mm. long. Flowers greenish. Sepals 5, 1–3 mm. long. Stamens 5, sometimes 3 or 10. Carpels 3, stigmas 3, short. Seeds brown, compressed. Fig. 5/12–13, p. 14.

var. **cerviana**

Basal leaves linear ; peduncles usually lacking. Fig. 5/12–13, p. 14.

UGANDA. Teso District : Serere, Nov.–Dec. 1931, *Chandler* 90 ! ; Busoga District : Central Forest Reserve, 16 km. N. of Jinja Mutai, 30 Oct. 1952, *G. H. S. Wood* 494 !
KENYA. Machakos, 12 Oct. 1947, *Bogdan* 1295 !
TANGANYIKA. Shinyanga District : Old Shinyanga, 2 Apr. 1932, *B. D. Burtt* 3722 ! ; Mpwapwa, 9 Apr. 1932, *Hornby* 457 !
DISTR. **U**1, 3–4 ; **K**1, 4, 6 ; **T**1, 2, 4–5 ; widespread in tropics and subtropics of the Old World ; introduced into America
HAB. Weed of roadsides and of cultivated and waste places and dry river-beds ; 400–1650 m.

SYN. *Pharnaceum cerviana* L., Sp. Pl. : 272 (1753)
Mollugo tenuissima Peter in Abh. Ges. Wiss. Göttingen, n.f. 13 : 254 (1928) ; F.D.O.-A. 2, Descr. : 28, t. 35, fig. 1 (1932) ; F.D.O.-A. 2 : 260 (1938). Type : Tanganyika : Pare District, N. of Buiko, *Peter* 10425 (B, holo.†)

var. **spathulifolia** *Fenzl* in Ann. Wien. Mus. 1 : 379 (1836). Type : India, *Wight* (K–W, isolecto. !)

Basal leaves spathulate ; peduncles usually present.

UGANDA. West Nile District : 4 Jan. 1906, *Dawe* 490 !
KENYA. Turkana District : Lake Rudolf, Ferguson's Gulf, 15 May 1953, *Padwa* 159 ! ; Kilifi District : Kibarani–Kilifi, 5 May 1949, *Jeffery* K607 !
TANGANYIKA. Pare District : Ngulu, May 1928, *Haarer* 1299 ! ; Tanga, Sept. 1954, *Semsei* 1785 ! ; Lindi District : about 9·5 km. S. of R. Mbemkuru, 7 Dec. 1955, *Milne-Redhead & Taylor* 7477 !
DISTR. **U**1 ; **K**2, 4, 7 ; **T**3, 6–8 ; widespread in the Old World Tropics
HAB. Weed of roadsides and of cultivated places ; 0–750 m.

SYN. *Pharnaceum umbellatum* Forsk., Fl. Aegypt.-Arab.: 58 (1775). Type : Arabia, *Forskål* (C, holo.)
Mollugo umbellata (Forsk.) Ser. in DC., Prodr. 1 : 393 (1824)
M. spathulifolia (Fenzl) Dinter in F.R. 19 : 236 (1923)

These two varieties of this species are here maintained, as they are quite distinct morphologically and seem to have rather different geographical and altitudinal distributions in our area. They are, however, perhaps less distinct in other parts of their range. Collectors should note if at any time the two varieties are seen growing together, and, if so, if any intermediates occur.

2. **M. nudicaulis** *Lam.*, Encycl. 4 : 234 (1797) ; F.T.A. 2 : 591 (1871) ; F.W.T.A. 1 : 114 (1927) ; ed. 2, 1 : 134 (1954) ; F.D.O.-A. 2 : 258 (1938) ; Andrews, Fl. Pl. Anglo-Egypt. Sudan 1 : 94 (1950) ; F.C.B. 2 : 110 (1951). Type : Mauritius, *Commerson* (?P, holo.)

A glabrous herb with a rosette of basal leaves and erect or ascending cymosely-branched leafless inflorescences, 2·5–35 cm. high. Leaves all radical, oblanceolate or obovate, entire, rounded at apex, narrowed to a sessile or subpetiolate base, 9–40 mm. long, 2–27 mm. wide. Inflorescences of several (rarely 1, sometimes many) forking branches arising from the crown. Branches leafless, bearing at the nodes only inconspicuous membranous brownish lanceolate bracts about 2 mm. long. Flowers greenish, pedicellate, solitary. Pedicels 1–1·5 mm. long. Sepals 5 ; stamens 3–5, hypogynous. Carpels 3, styles 3, short. Seeds compressed, ovate, granular.

UGANDA. West Nile District : Maracha Rest Camp, 6 Aug. 1953, *Chancellor* 122 ! ; Kigezi District : Nyamugoye, Kinziki, Mar. 1951, *Purseglove* 3582 ! ; Busoga District : just NE. of Buwongo Hill, about 5 km. SW. of Mukuta, 17 Apr. 1953, *G. H. S. Wood* 724 !
KENYA. West Suk District : Sigor, 80 km. N. of Kapenguria, Weinei Valley, 13 Nov. 1953, *Bogdan* 3835 ! ; Nairobi District : Njiro Farm, 19 km. E. of Nairobi, 19 June 1951, *Bogdan* 3069 ! Teita District : Voi, Mutanda Rocks, 8 Feb. 1953, *Bally* 8797 !
TANGANYIKA. Pare District : Kisangara, May 1928, *Haarer* 1300 ! ; Lushoto District : about 5 km. NW. of Mombo, 29 Apr. 1953, *Drummond & Hemsley* 2274 ! ; Morogoro District : Turiani, 3 Nov. 1953, *Semsei* 1425 !
DISTR. **U**1–4 ; **K**1–5, 7 ; **T**1–6, 8 ; pantropical
HAB. A weed of roadsides and of open and waste places and cultivated ground ; 0–1800 m.

SYN. *Pharnaceum spathulatum* Sw., Fl. Ind. Occ. 1 : 568 (1797). Type : Jamaica, *Hermann* (BM–SL, lecto. !)
P. bellidifolium Poir., Encycl. 5 : 262 (1804). Type : probably from French Guiana, (?P, holo.)
Mollugo bellidifolia (Poir.) Ser. in DC., Prodr. 1 : 391 (1824)

7. HYPERTELIS

Fenzl in Ann. Wien. Mus. 2 : 261 (1839)

[*Pharnaceum* sensu auct., *non* L., Sp. Pl.: 272 (1753) sensu stricto]

Hyperstelis Pax in E. & P. Pf. III. lb : 40 (1889)

Annual or perennial sometimes subwoody herbs. Leaves alternate or whorled, often crowded, linear, entire, succulent, glaucous, stipulate. Stipules persistent, membranous, entire, adnate to the leaf-base, sometimes stem-clasping. Inflorescences axillary, long-pedunculate, simply umbelliform. Sepals 5, free. Stamens 3–15, alternate with sepals, singly or (when more than 5) in pairs or fascicles, hypogynous. Disc absent. Ovary superior, syncarpous, of 3–5 carpels, 3–5-locular ; ovules numerous, axile ; stigmas 3–5, short.

9 species, mostly South African, but one extending to tropical Africa and Madagascar and one endemic in St. Helena.

H. bowkeriana *Sond.* in Harv. & Sond., Fl. Cap. 1 : 145 (1860). Type : South Africa, Cape Province, *Bowker* (K, holo. !)

A perennial shrubby herb up to 30 cm. high with ascending procumbent or creeping branches 5–40 cm. long. Leaves alternate or subopposite, and also in axillary fascicles on short side-branches at the nodes, linear, succulent, glaucous, subcylindrical, often with wart-like glands, 12–60 mm. long, 0·5–1·0 mm. wide. Stipules membranous, triangular, somewhat divaricate, arising from the expanded membranous margins of the leaf-bases. Pedicels 3–9 cm. long, usually smooth but sometimes (like the pedicels and calyces) with a few obscure glandular warts. Stamens 5. Fig. 6.

KENYA. Machakos District : Athi River Stn., 29 Aug. 1947, *Bally* 5247 ! ; Lamu, W. side of town, 15 Feb. 1956, *Greenway & Rawlins* 8912 !
DISTR. **K**4, 7 ; Ethiopia southwards to Cape Province of South Africa, also in Madagascar
HAB. Saline sandy maritime and inland soils in open situations ; 0–1600 m.

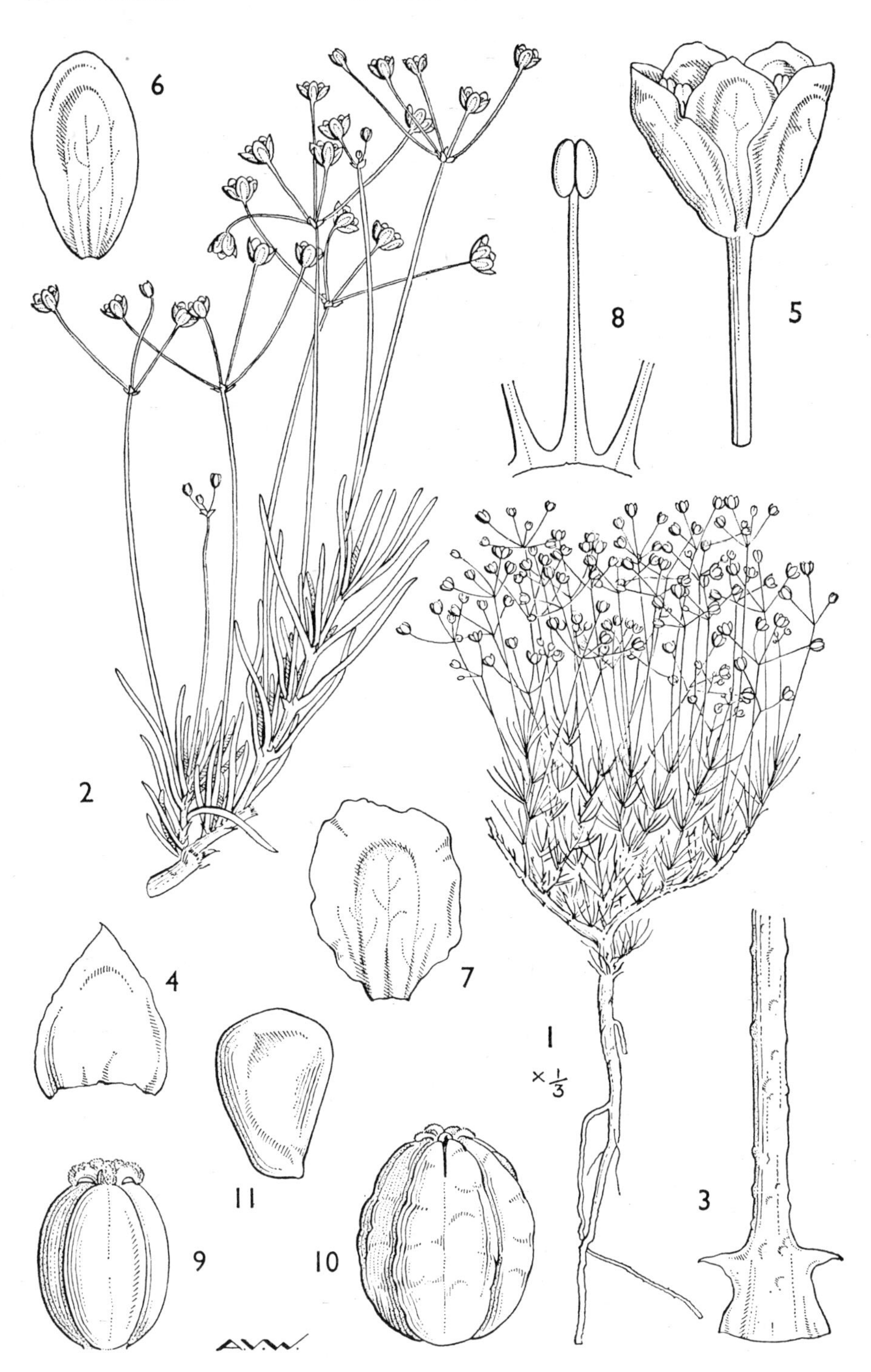

FIG. 6. *HYPERTELIS BOWKERIANA*, from *Greenway & Rawlins* 8912—**1,** habit, × ⅓ ; **2,** inflorescences, × 1 ; **3,** leaf-base, × 4 ; **4,** bract, × 12 ; **5,** flower, × 8 ; **6,** outer sepal, × 8 ; **7,** inner sepal, × 8 ; **8,** stamen, × 12 ; **9,** ovary, × 12 ; **10,** fruit, × 10 ; **11,** seed, × 40.

SYN. *Pharnaceum suffruticosum* Bak. in J.L.S. 20 : 151 (1883). Type : Madagascar, Ambongo, *Pervillé* 647 (K, holo. !)
Mollugo suffruticosa Peter in Abh. Ges. Wiss. Göttingen, n.f. 13 (2) : 54 (1928) ; F.D.O.-A. 2, Descr.: 29, t. 34. (1932) ; F.D.O.-A. 2 : 259 (1938). Type : Tanganyika : Pare District, NW. of Buiko towards Hedaru, *Peter* 41190 (B, holo. †)
M. suffruticosa Peter forma *annua* Peter, F.D.O.-A. 2, Descr. : 29 (1932) ; F.D.O.-A. 2 : 259 (1938). Type : Lushoto District : Lake Manga, *Peter* 10864 (B, holo.†)
Hypertelis suffruticosa (Bak.) Adamson in J. S. Afr. Bot. 24 : 55 (1958)

8. **SESUVIUM**

L., Syst. Nat., ed. 10 : 1058 (1759)

Erect, decumbent or creeping succulent annual or perennial herbs or sub-shrubs. Leaves opposite, subopposite, or alternate, sessile or petiolate, linear, lanceolate, oblong, elliptic or ovate, entire, succulent. Stipules absent. Flowers axillary, solitary, sessile or pedunculate, bibracteolate, hermaphrodite, regular. Calyx gamosepalous, 5-partite ; calyx-lobes triangular, dorsally apiculate, green outside, coloured inside ; calyx-tube short, obconical. Stamens 5-many, free or connate at the base, inserted in the mouth of the calyx-tube. Carpels 2–5, superior, united ; ovary 2–5-locular ; styles 2–5 ; ovules many ; placentation axile. Fruit a circumscissile capsule, the lid remaining whole. Seeds several to many, black.

6 species, one a pantropical littoral, one endemic in the Galapagos Is., the others mostly confined to the Angolan region.

Flowers pedicellate, the pedicel 3–15 mm. long ; stems smooth, rooting at the nodes ; plant perennial . 1. *S. portulacastrum*
Flowers sessile ; stems usually rough-warty, not rooting at the nodes ; plant annual . . . 2. *S. sesuvioïdes*

1. **S. portulacastrum** (*L.*) *L.*, Syst. Nat., ed. 10: 1058 (1759) ; F.T.A. 2 : 585 (1871) ; F.W.T.A. 1 : 115 (1927) ; ed. 2, 1 : 135 (1954) ; F.D.O.-A. 2 : 263 (1938). Type : Curaçao, *Hermann* (BM–SL, typo. !)

A suberect prostrate or creeping glabrous succulent perennial herb. Stems thick, smooth, rooting at the nodes. Leaves opposite, oblanceolate, fleshy, entire, rounded at the apex, narrowed gradually below into a petiole, the upper surface flat, the lower convex, the base scarious-expanded, stem-clasping and connate with that of the opposing leaf, 10–65 mm. long (including petiole), 1–14 mm. wide. Flowers solitary, 7–12 mm. long, pedicellate, the pedicels thickened upwards, 3–15 mm. long. Calyx-lobes unequal, triangular, acute, each with a fleshy dorsal apiculus about 1·5 mm. long just below the apex, green outside, pink, red or purplish inside. Calyx-tube short, about one third the length of the lobes. Stamens many, free. Ovary of 3 (rarely 4) carpels, 3-(rarely 2- or 4-) celled ; styles 3–4. Seeds black, smooth. Fig. 7.

KENYA. Kwale District : Vanga, Nov. 1929, *R. M. Graham* 2221 !
TANGANYIKA. Tanga District : N. end of Boma peninsula, near Bomandari, 3 Aug. 1953, *Drummond & Hemsley* 3625 ! ; Rufiji District : Mafia Is., 31 Mar. 1933, *Wallace* 822 !
ZANZIBAR. Zanzibar Is., Mbweni, 4 Feb. 1929, *Greenway* 1335 !
DISTR. **K**7 ; **T**3, 6 ; Z ; pantropical
HAB. Maritime shores at and about high water-level

SYN. *Portulaca portulacastrum* L., Sp. Pl. : 446 (1753)

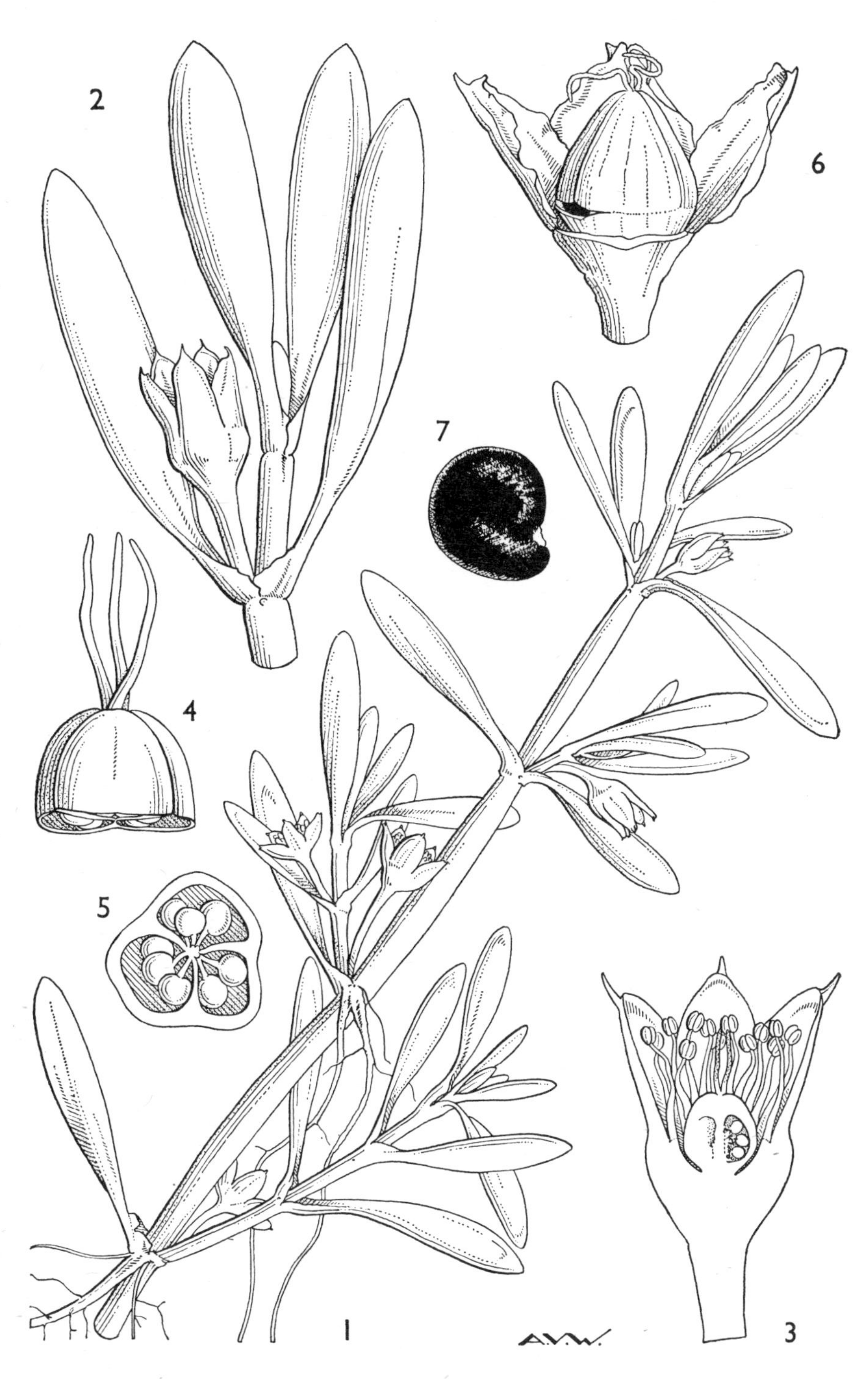

FIG. 7. *SESUVIUM PORTULACASTRUM*—**1**, habit, × 1 ; **2**, flower, × 2 ; **3**, flower, in longitudinal median section, × 4 ; **4**, operculum, × 8 ; **5**, ovary, in transverse section, × 8 ; **6**, fruit in process of dehiscing, × 4 ; **7**, seed, × 12. 1, 3–7, from *Drummond & Hemsley* 3625 ; 2, from *Faulkner* 2108.

2. **S. sesuvioïdes** (*Fenzl*) *Verdc.* in K.B. 12 : 349 (1957). Type : South Africa, Orange River, *Drège* 2938 (W, holo. †, K, iso. !)

A prostrate or straggling succulent annual or ? short-lived perennial herb. Stems usually rough with warty outgrowths, not rooting at the nodes, 2–30 cm. long or more. Leaves alternate (by reduction), subopposite, or opposite, narrowly elliptic to broadly ovate, petiolate ; the blade 7–20 mm. long, 2–11 mm. wide ; the petiole 3–10 mm. long, scarious-winged at the base. Flowers solitary, 5–14 mm. long, sessile. Calyx-lobes equal, narrowly triangular, acute, each with a dorsal apiculus 1–2 mm. long from just below the apex, pink inside, green outside. Stamens 5-many, free. Ovary of 2 carpels, 2-locular ; styles 2. Seeds about 2–10 in each loculus, rugose, black.

KENYA. Turkana District : 16 km. N. of Kagatet, 5 July 1954, *Hemming* 297 !
TANGANYIKA. Southern Province, without locality or date, *Busse* 1071 !
DISTR. **K**2 ; **T**8 ; India ; NE. tropical Africa and the Sudan westwards to the Cape Verde Is. ; Nyasaland ; Southern Rhodesia and Bechuanaland ; Angola, South West Africa, Cape Province
HAB. Dry open sandy places

SYN. *Diplochonium sesuvioïdes* Fenzl, Nov. Stirp. Dec. 7 : 58 (1839) & in Ann. Wien. Mus. 2 : 292 (1839)
Trianthema hydaspica Edgew. in J.L.S. 6 : 203 (1862). Type : India, *Edgeworth* 3019 (K, lecto. !)
Sesuvium digynum Oliv., F.T.A. 2 : 586 (1871). Type : Angola, Moçâmedes, *Welwitsch* 2392 (K, lecto. !)
Trianthema polysperma Oliv., F.T.A. 2 : 588 (1871). F.D.O.-A. 2 : 263 (1938) ; Andrews, Fl. Anglo-Egypt. Sudan 1 : 96 (1950) ; F.W.T.A., ed. 2, 1 : 136 (1954). Type : Sudan, *Kotschy* 68 (K, holo. !)
Halimus sesuvioïdes (Fenzl) O. Ktze., Rev. Gen. 2 : 263 (1891)
Trianthema nyassica Bak. in K.B. 1897 : 268 (1897). Type : Nyasaland, Lake Nyasa, Monkey Bay, *Whyte* (K, holo. !)
Sesuvium digynum Oliv. var. *angustifolium* Schinz in Bull. Herb. Boiss. 5, app. 3: 74 (1897). Type : Angola, Benguela, *Wawra* 291 (W, holo.†)
S. hoepfnerianum Schinz in Bull. Herb. Boiss 5, app. 3 : 75 (1897). Type : South West Africa, *Höpfner* 25 (Z, holo.)
S. hoepfnerianum Schinz var. *brevifolium* Schinz in Bull. Herb. Boiss. 5, app. 3 : 75 (1897). Type : South West Africa, *Steingrover* 38 (Z, holo.)
Halimum sesuvioïdes (Fenzl) Hiern var. *reduplicatum* Hiern, Cat. Welw. Afr. Pl. 1 : 414 (1898). Type : Angola, Moçâmedes, *Welwitsch* 2391 (BM, lecto. !, K, isolecto. !)
H. sesuvioïdes (Fenzl) Hiern var. *welwitschii* Hiern, Cat. Welw. Afr. Pl. 1 : 414 (1898). Type : Angola, Moçâmedes, *Welwitsch* 2392 (BM, lecto. !, K, isolecto. !)
H. sesuvioïdes (Fenzl) Hiern var. *augustifolium* (Schinz) Hiern, Cat. Welw. Afr. Pl. 1 : 414 (1898)
Trianthema salaria Bremek. in Ann. Transv. Mus. 15 : 239 (1953). Type : South Africa, Transvaal, *Bremekamp & Schweickerdt* 232 (PRE, holo.)

This species as here circumscribed shows considerable variation, especially in internode length, flower size, and number of stamens. Most of the variation is found in plants from South West Africa, southern Angola and adjoining areas of Rhodesia, Bechuanaland and Cape Province ; only in this region do the typical large-flowered plants with many stamens occur. Plants from the rest of Africa (including our area) and India appear to be more uniform with few (usually 5) stamens and generally smaller flowers. Further study of South West African material is needed to establish if in fact there is sufficient discontinuity in variation for the delimitation of subordinate taxa.

9. TRIANTHEMA

L., Sp. Pl. : 223 (1753) & Gen. Pl., ed. 5 : 105 (1754)

Procumbent diffuse glabrous, papulose, or pubescent succulent mostly annual herbs. Leaves opposite, those of a pair often unequal, linear to almost orbicular, entire, petiolate or sessile. Leaf-bases membranous-dilated, often connate in pairs and with small stipuliform lobes. Flowers axillary, solitary, fascicled or glomerulate, free or connate, sessile or stalked. Calyx gamosepalous, 5-lobed ; lobes usually with a distinct subapical dorsal mucro. Stamens 5–many, episepalous, when definite alternate with the calyx-lobes. Ovary of one carpel, unilocular, apex truncate or depressed around the style ; ovules and seeds 2–many ; placentation parietal.

About 9 species, one a pantropical weed originally American, one in Argentina, the rest in tropical Africa, Asia and Australia.

Flowers solitary, or if a few together, then 8–15 mm. long:
  Plant almost completely glabrous ; flowers more or less hidden by the sheathing leaf-bases ; calyx-tube adnate to the petiole . . . . 1. *T. portulacastrum*
  Plant densely hispid-pilose ; flowers not at all hidden by the leaf-bases ; calyx-tube quite free from the petiole . . . . . . 2. *T. ceratosepala*
Flowers glomerulate, or if solitary or subsolitary, then 2–4 mm. long :
  Calyx-lobes 0·5–1 mm. long, about as long as broad, when dry held upright or horizontal across the mouth of the calyx-tube, but not tightly pressed together along their whole lengths ; dorsal subapical mucro obsolescent, 0·3 mm. long or less, or represented merely by a thickening ; calyces 2–4 mm. long, basally free, never forming a subspherical compound mass when in fruit . . . . 3. *T. triquetra*
  Calyx-lobes 1·5–2 mm. long, longer than broad, when dry held upright and adpressed together, forming a cone ; dorsal mucro obvious, 0·5–1·5 mm. long ; calyces 3–6 mm. long, basally connate, often forming a subspherical compound mass when in fruit . . . 4. *T. salsoloïdes*

1. **T. portulacastrum** *L.*, Sp. Pl.: 223 (1753) ; F.W.T.A. 1 : 115 (1927) ; ed. 2, 1 : 136 (1954) ; F.C.B. 2 : 116 (1951). Type : South America, Curaçao, *Hermann* (BM–SL, typolecto.!)

A somewhat succulent subglabrous annual herb ; stems procumbent or ascending, spreading, glabrous or sparsely hairy, up to 50 cm. long or more. Leaves opposite, one of a pair much smaller than the other, petiolate, stipulate ; blade obovate or broadly so, entire, obtuse, rounded or retuse, sometimes slightly apiculate at apex, cuneate at base, glabrous or sparsely hairy on the midrib below, 4–50 mm. long, 4–45 mm. broad. Petiole 2–25 mm. long, distinct, sparsely hairy, expanded into a sheathing membranous base connate with that of the opposing leaf. Stipules narrowly triangular, acuminate, up to 3 mm. long. Flowers axillary, solitary, partly hidden by the sheathing leaf-bases, hermaphrodite, pinkish or white, 4–5 mm. long. Stamens 10–20, inserted on the calyx-tube. Calyx-lobes obtuse, with a dorsal apiculus. Ovary truncate, bilobed ; seeds 3–12.

Kenya. Mombasa, 24–25 Apr. 1927, *Linder* 2634 !
Tanganyika. Tanga District : Tengeni, 26 Nov. 1939, *Greenway* 5905 !
Distr. **K**7 ; **T**3, 6, 8 ; pantropical
Hab. Weed of cultivated ground ; 0–250 m.

Syn. *Trianthema monogyna* L., Mant. 1 : 69 (1767) ; F.T.A. 2 : 587 (1871) ; F.D.O.-A. 2 : 261 (1938), *nom. illegit.* Type : as *T. portulacastrum*

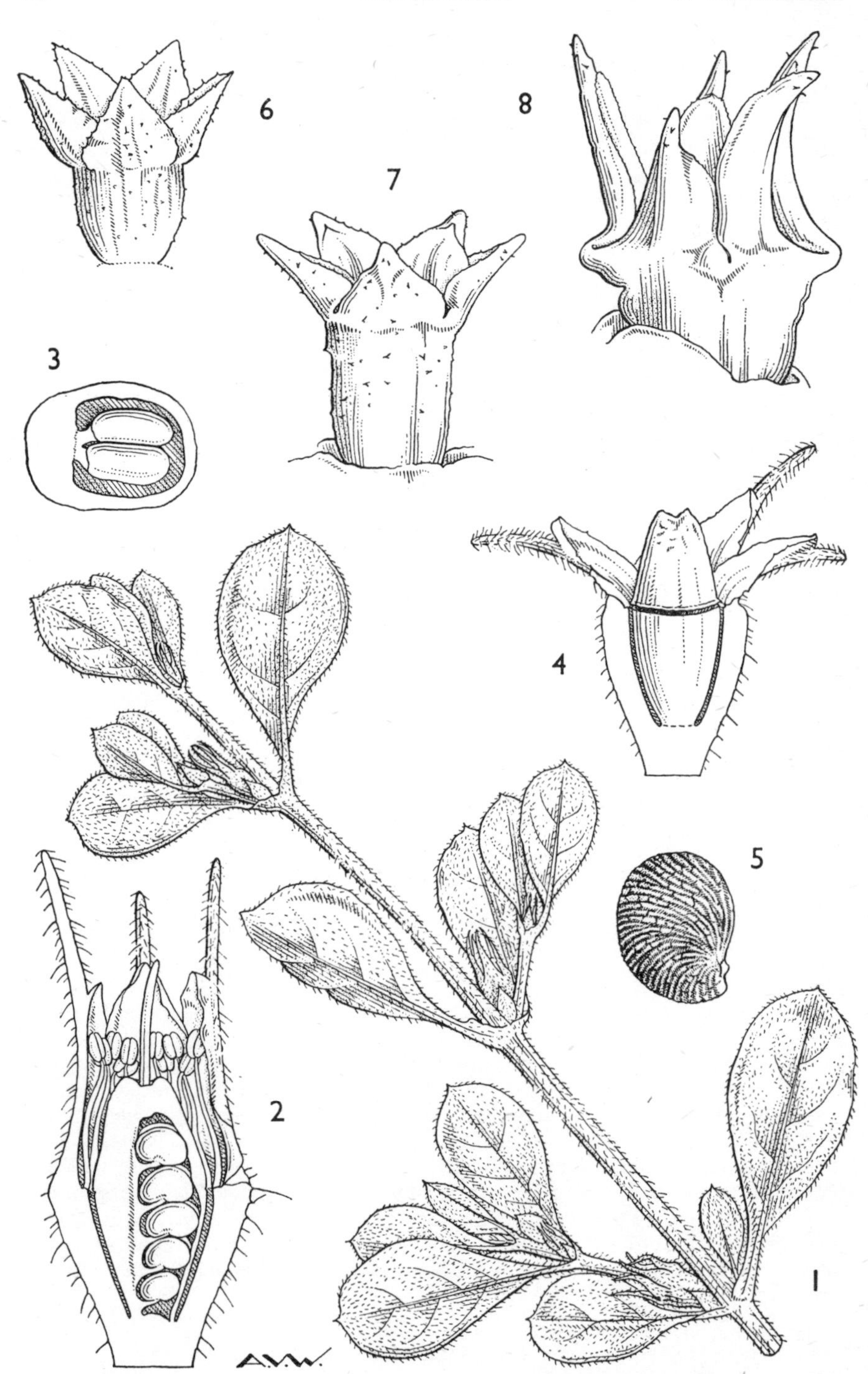

FIG. 8. *TRIANTHEMA CERATOSEPALA*—**1,** habit, × 1 ; **2,** flower, in longitudinal median section, × 6 ; **3,** ovary, in transverse section, × 12 ; **4,** fruit in process of dehiscing, × 4 ; **5,** seed, × 8 ; *T. TRIQUETRA* subsp. *TRIQUETRA* var. *TRIQUETRA*—**6,** flower, × 12 ; var. *SANGUINEA*—**7,** flower, × 12 ; *T. SALSOLOÏDES*—**8,** flower, × 12. 1–5, from *Bally* 8130 ; 6, from *Bogdan* 4226 ; 7, from *Bally* 3546 ; 8, from *Bally* 11587.

2. **T. ceratosepala** *Volkens & Irmsch.* in E.J. 48 : 497 (1912). Type : Tanganyika, Moshi District, Kahe, *Volkens* 2219 (B, holo.)

A succulent densely-pilose spreading prostrate perennial herb or subshrub. Branches retrorsely pilose, up to 40 cm. or more in length. Leaves opposite, one of each pair somewhat smaller than the other, petiolate, entire, retrorsely adpressed-hairy ; blade obovate, oblanceolate or elliptic, subacute, obtuse or rounded at the apex, 12–35 mm. long, 4–18 mm. broad. Petiole hairy, expanded at the base, flattened, 3–10 mm. long, the bases often persistent after the fall of the leaves. Flowers usually solitary, axillary, magenta, sessile, 8–15 mm. long. Calyx-lobes with bright purple petaloid margins, somewhat cucullate at the apex, each with a dorsal apiculus about 3 mm. long. Stamens 20–30, inserted on the calyx-tube. Ovary truncate, bilobed. Seeds about 6. Fig. 8/1–5.

KENYA. Masai District : Amboseli Reserve, N. of Kilimanjaro, 12 Sept. 1954, *Bally* 9868 !
TANGANYIKA. Moshi District : Himo Plains, N. of N. Pare Mts., 2 Apr. 1952, *Bally* 8130 !
DISTR. **K**6 ; **T**2, 3 ; Ethiopia, Somaliland
HAB. On alkaline soils and in *Acacia* bushland ; 600–1200 m.

3. **T. triquetra** *Willd.* in Ges. Naturf. Fr. Berlin, Schr. 4 : 181 (1803). Type : India, *Rottler* (B–W, holo., K, iso. !, photo. !)

A more or less succulent procumbent or prostrate annual herb, much branched from the base. Branches radiating, terete, often tinged reddish and papulose, especially in the younger parts. Leaves linear, oblanceolate, oblanceolate-spathulate, obovate or almost orbicular, petiolate or subsessile ; blade succulent, oblong-elliptic to almost circular in cross-section, 1–22 mm. long, 1–4 mm. broad. Petiole 0–5 mm. long ; leaf-base expanded, membranous, slightly 2-lobed, stem-clasping, sometimes persistent on the stem. Flowers axillary, glomerulate (usually 2–6, sometimes more, rarely solitary), basally free from one another, sessile or shortly stalked, 2–4 mm. long. Calyx-segments 5, triangular, each with a subapical dorsal thickening, at its largest forming an obsolescent mucro up to 0·3 mm. long, usually glabrous, sometimes papulose-pubescent. Stamens 5, alternate with the calyx-lobes. Ovary obconical, turbinate, depressed above centrally around the short style. Ovules and seeds 2, superposed, sometimes 1 by abortion of the other. Fig. 8/6–7.

subsp. **triquetra**

Leaves linear to narrowly oblanceolate-obovate, subsessile or petiolate ; flowers sessile ; calyx-lobes about as broad across the base as long. Fig. 8/6–7.

var. **triquetra**

Calyx 2–3 mm. long, the lobes not glassy-papillate pubescent on the backs and edges, the latter thus appearing smooth ; stems and flowers usually greenish. Fig. 8/6.

UGANDA. Karamoja District : Kantaku Enclosure, Sept. 1958. *Wilson* 625 !
KENYA. Baringo District : S. shore of Lake Baringo, 21 Aug. 1956, *Bogdan* 4226 ! ; 16 km. S. of Lake Baringo, 5 Jan. 1959, *Bogdan* 4739 !, 4742 ! Kwale District : between Samburu and Mackinnon Road, 1 Apr. 1953, *Drummond & Hemsley* 4090 !
TANGANYIKA. Mwanza, *R. L. Davis* 116 ! ; Mbulu District : Tarangire River, 17 May 1958, *Makinda* 391 ! ; Lushoto District, 8 km. SE. of Mkomazi, 30 Apr. 1953, *Drummond & Hemsley* 2299 !
DISTR. **U**1 ; **K**1, 3, 7 ; **T**1–3, 7 ; Sudan, Ethiopia, Somaliland, Arabia, India, Malaysia, Australia ; in South and South West Africa represented by the subsp. *parvifolia* (Sond.) Jeffrey
HAB. Open habitats in grassland, bushland and woodland and on maritime and inland saline sands and silts ; 0–1100 m.

SYN. *Trianthema sedifolia* Vis., Pl. Aegypt. & Nub. : 19, t. 3, fig. 1 (1836) ; F.T.A. 2 : 588 (1871) ; Andrews, Fl. Pl. Anglo-Egypt. Sudan 1 : 96 (1950) ; F.W.T.A., ed 2, 1 : 136 (1954). Type : Sudan, Khartoum, *Brocchi* (BASSA, holo.)
[*T. crystallina* sensu auct., & sensu Peter, F.D.O.-A. 2 : 261 (1938), *non* (Forsk.) Vahl, Symb. Bot. 1 : 32 (1790)]
T. glandulosa Peter, F.D.O.-A. 2, Descr. : 30, t. 36, fig. 1 (1932) ; F.D.O.-A. 2 : 262 (1938). Type : Tanganyika, Masai District, Emugur Belekj, *Peter* 42741b (B, holo.†)

Plants from saline soils tend to have thicker more fleshy stems and leaves and larger, tougher membranous leaf-bases which tend to persist on the stem. In extreme cases these tendencies give rise to plants looking very different from the typical ones ; such plants are characteristic of the maritime sands north of Lamu on the Kenya coast ; an example is :—Kenya, Lamu District, Takwa strand, 1956, *Rawlins* 54 !

var. **sanguinea** (*Volkens & Irmsch.*) *Jeffrey* in K.B. 14 : 237 (1960). Type : Tanganyika : Moshi District : between Himo and Pangani R., *Volkens* 458 (B, holo. !, K, photo. !)

Calyx 2–4 mm. long, the lobes glassy papillate-pubescent on the backs and edges, the latter thus appearing rough ; stems and flowers tinged reddish. Fig. 8/7, p. 24.

KENYA. Masai District : Amboseli Reserve, 14 Sept. 1954, *Bally* 9866 ! & foot of Ol Lorgosailic, 27 May 1956, *Bally* 10559 !
TANGANYIKA. Moshi District : between Himo and the Pangani R., 5 July 1893, *Volkens* 458 !
DISTR. **K**6, **T**2 ; not known elsewhere
HAB. Alkaline, alluvial, and volcanic soils, 700–1200 m.

SYN. *Trianthema sanguinea* Volkens & Irmsch. in E.J. 48 : 498 (1912)
T. nigricans Peter, F.D.O.-A. 2, Descr. : 30, t. 36, f. 2 (1932) ; F.D.O.-A. 2 : 262 (1938). Type : Tanganyika, Masai District : Emugur Belekj, *Peter* 42741c. (B, holo.†)

4. **T. salsoloïdes** *Oliv.* in F.T.A. 2 : 588 (1871) ; F.D.O.-A.2 : 262 (1938) ; Andrews, Fl. Pl. Anglo-Egypt. Sudan 1 : 96 (1950). Type : Sudan, *Kotschy* 137 (K, holo. !)

A succulent smooth or more usually papillose much-branched annual herb ; stems prostrate, ascending or erect, radiating from the crown of the taproot, sometimes tinged reddish, 10–30 cm. or more in length. Leaves linear to narrowly obovate or oblong-ovate, subsessile, 9–35 mm. long, 1–9 mm. broad ; petiole 0–3 mm. long ; leaf-base expanded, membranous, with 2 stipular lobes. Flowers 3–6 mm. long, sessile, axillary, glomerulate, 3–12 or more, rarely solitary, the calyces basally connate and often forming a compact compound subspherical mass in the fruiting state. Calyx-lobes longer than broad, smooth or papillose, each with a distinct dorsal subapical fleshy mucro 0·5–1·5 mm. long. Calyx-tube obconical, bearing 5 more or less prominent tubercle-like projections at the base of the sinuses between the lobes. Stamens 5. Ovary obconical, turbinate, depressed above around the style. Ovules and seeds 2, sometimes one by abortion of the other. Fig. 8/8, p. 24.

TANGANYIKA. Shinyanga District : Seseku, 10 June 1931, *B. D. Burtt* 2521 ! ; Mbulu District: N. end of Lake Eyasi, 23 July 1957, *Bally* 11587 ! ; Pangani R., July 1893, *Volkens* 451 !
DISTR. **T**1–2, 4–5 ; Sudan, Ethiopia, Rhodesia, Transvaal ; also India
HAB. Grassland and bushland, 900–1260 m.

SYN. *Trianthema transvaalensis* Schinz in Vierteljahrsschr. Nat. Ges. Zürich 60 : 396 (1915). Type : South Africa, Transvaal, *Schlechter* 4876 (Z, holo., K, iso. !)
T. crystallina (Forsk.) Vahl var. *oblongifolia* Gamble, Fl. Madras 1 : 551 (1918). Type : India, *Bourne* (K, holo. !)
T. multiflora Peter, F.D.O.-A. 2, Descr. : 29, t. 35, f. 2 (1932) ; F.D.O.-A. 2 : 262 (1938). Type : Tanganyika, Masai District, Emugur Belekj, *Peter* 42741a (B, holo. †)

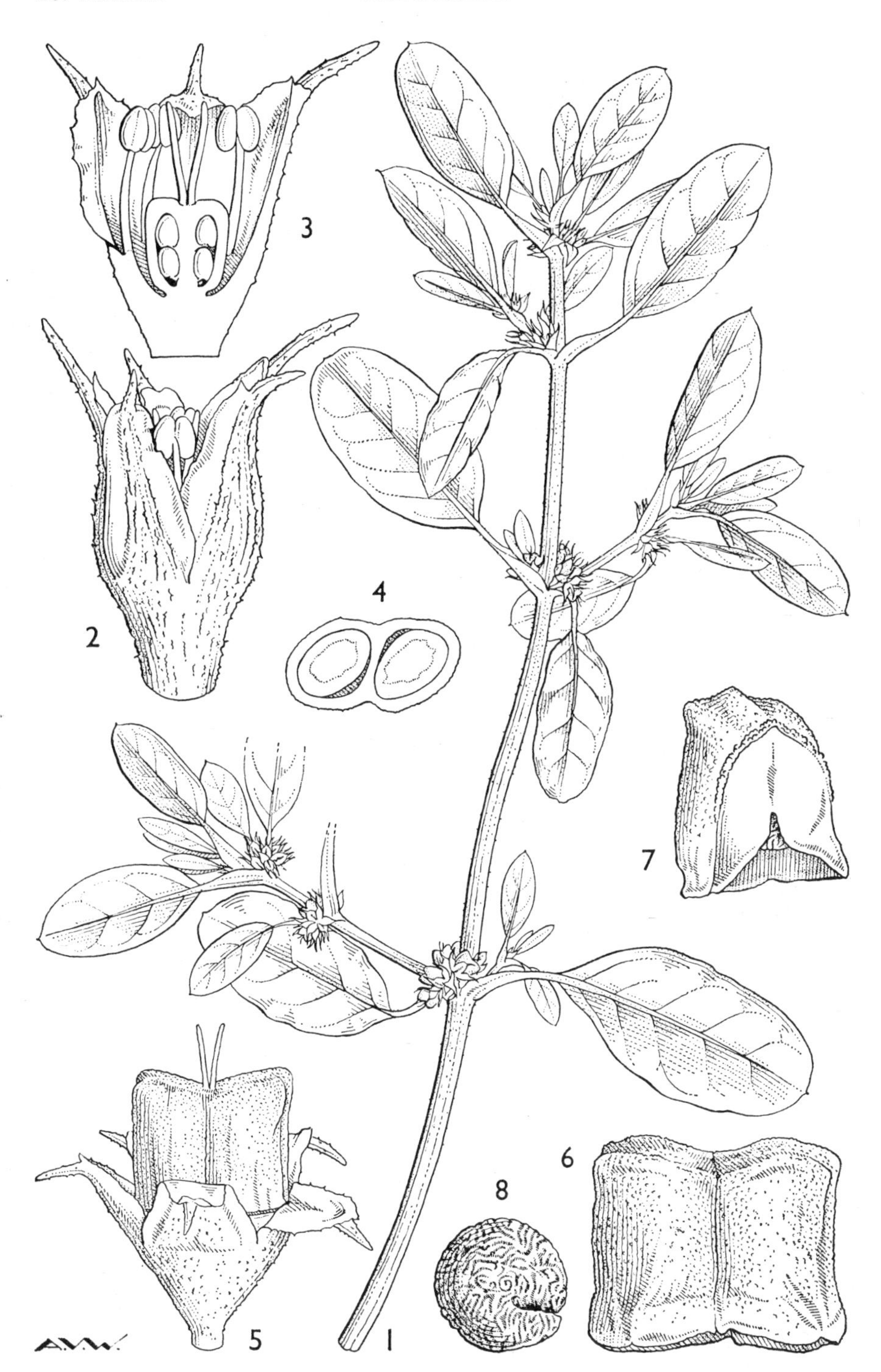

FIG. 9. *ZALEYA PENTANDRA*, from *Drummond & Hemsley* 1046—**1,** habit, × 1; **2,** flower, × 12; **3,** flower, in longitudinal median section, × 12; **4,** ovary, in transverse section, × 16; **5,** fruit, × 8; **6,** operculum of fruit, × 12; **7,** valve of operculum showing commissural face, × 12; **8,** seed, × 10.

10. **ZALEYA**

Burm. f., Fl. Ind. : 110, t. 31 (1768)

Rocama Forsk., Fl. Aegypt.-Arab. : 71 (1775)

Usually prostrate annual and perennial herbs. Leaves opposite, petiolate, lanceolate or oblanceolate, narrowly to broadly elliptic or ovate, entire, slightly succulent. Stipules absent. Flowers axillary, glomerulate, subsessile, hermaphrodite, regular. Calyx gamosepalous, 5-lobed ; tube short ; lobes membranous-margined, coloured inside, green outside, and with a subapical dorsal mucro. Stamens 5–15, free, inserted in the upper part of the calyx-tube. Ovary superior, syncarpous, 2-celled, of 2 united carpels ; stigmas 2, free ; ovules 2 per cell, attached to the interlocular septum. Fruit a four-seeded capsule, dehiscing by means of a bivalved operculum, the valves usually separating.

About 6 closely related species in tropical Africa, Asia and Australia.

Z. pentandra (*L.*) *Jeffrey* in K.B. 14 : 238 (1960). Type : Arabia, cultivated at Uppsala from seed sent by Forskal (LINN, lecto.)

A spreading prostrate or procumbent, rarely erect, slightly succulent herb, sometimes subwoody at the base. Branches pubescent when young, becoming glabrous, 8–30 cm. long or more. Leaf-blade oblanceolate, narrowly elliptic, elliptic, or broadly so, obtuse to rounded at the apex, 8–38 mm. long, 2–21 mm. broad, glabrous or pubescent below or on the midrib only. Petiole membranous-winged and sheathing at the base, usually pubescent, 3–18 mm. long. Flowers 5–20 per glomerule, sessile or subsessile, greenish, tinged pink or crimson. Stamens 5. Seeds 1·7 mm. diameter, black, ribbed. Fig. 9, p. 27.

UGANDA. Karamoja District : Kangole, 20 May 1940, *A. S. Thomas* 3486 ! ; Ankole District : Ruizi R., 13 Nov. 1950, *Jarrett* 177 !
KENYA. Ravine District : 24 km. SE. of Eldama Ravine, 23 Aug. 1956, *Bogdan* 4251 ! ; Central Kavirondo District : Sukuri Is., Kisumu, July 1934, *Miss Napier & Fox in C.M.* 6641 ! ; Mombasa Is., 28 Jan. 1953, *Drummond & Hemsley* 1046 !
TANGANYIKA. Musoma District : Bunogi Area, Magungu R., 6 Nov. 1953, *Tanner* 1716 ! ; Lushoto District : Mazinde, 1 May 1953, *Drummond & Hemsley* 2333 ! ; Dodoma District : Mwitikira, 22 Aug. 1936, *Greenway* 812 !
DISTR. **U**1–2 ; **K**1–7 ; **T**1–3, 5–7 ; tropical Africa in the drier areas from Transvaal to Egypt and Senegal ; also in Arabia, Palestine and in Madagascar.
HAB. Open places in woodland, bushland and grassland ; also a weed of overgrazed and waste places, roadsides, and cultivated ground ; and on alkaline soils, 0–2000 m.

SYN. *Trianthema pentandra* L., Mant. 1 : 70 (1767) ; F.T.A. 2 : 588 (1871) ; F.W.T.A. 1 : 115 (1927) ; ed. 2, 1 : 136 (1954) ; F.D.O.-A. 2 : 262 (1938) ; Andrews, Fl. Pl. Anglo-Egypt. Sudan 1 : 96, fig. 61. (1950) ; F.C.B. 2 : 116 (1951)
Rocama prostrata Forsk., Fl. Aegypt.-Arab. : CVIII (nomen) & 71 (descr.) (1775). Type : Arabia, *Forskål* (C, holo.)
Limeum kenyense Suesseng. in Mitt. Bot. Staatss. München 2 : 46 (1950). Type : Kenya, Turkana District, *Martin* 17 (K, holo. !)
Trianthema redimita Melville in K.B. 7 : 268 (1952). Type : Kenya, Northern Frontier Province, Mudo near Wajir, *Dale* K700 (K, holo. !)

11. **AÏZOÖN**

L., Sp. Pl. : 488 (1753) & Gen. Pl., ed. 5 : 216 (1754) ; Adamson in J. S. Afr. Bot. 25 : 29 (1959)

Often succulent herbs or sometimes subshrubs. Leaves various, glabrous, papillose or hairy, alternate or opposite, exstipulate. Flowers hermaphrodite, solitary or in groups, sessile or pedunculate, often borne in the forkings of the stem. Calyx gamosepalous ; tube short ; lobes 4–5. Stamens many, inserted on the calyx-tube, often in 2 or more rows or in fascicles alternate with the calyx-lobes. Ovary superior, of 4–10 carpels, 4–10-locular ; ovules 2–many per loculus, pendulous, axillary ; styles as many as carpels, free. Fruit a loculicidal capsule, with as many or twice as many valves as carpels.

About 25 species in three distinct centres, the Mediterranean, South Africa and Australia, each with its distinct species, except for the following which is both Mediterranean and South African.

A. canariënse *L.*, Sp. Pl. : 488 (1753) ; F.T.A. 2 : 583 (1871) ; F.W.T.A. 1 : 115 (1927) ; Andrews, Fl. Pl. Anglo-Egypt. Sudan 1 : 95 (1950). Type : Canary Is. (LINN, lecto.)

A prostrate often rather thick-stemmed and tough annual or perennial herb. Stem pilose, often also finely papillose, 1–5 mm. diameter, 0·5–30 cm. long or more. Leaves alternate, petiolate ; blade suborbicular to oblanceolate-obovate, rounded obtuse or bluntly subacuminate at apex, decurrent into the petiole at the base, more or less pilose on both sides, entire, 6–40 mm. long, 2–40 mm. broad ; petiole 3–16 mm. long. Flowers solitary in leaf-axils or in the forks of the branches, sessile, often numerous. Calyx-lobes 5 (rarely 4) triangular, acute, yellowish inside, greenish or reddish and pilose outside. Stamens about 12–15, inserted on the calyx-tube in fascicles at the bases of the sinuses between the calyx-lobes. Styles short, deciduous. Fruit usually red or pink, pentagonal (rarely tetragonal) stelliform, depressed centrally, 5–8 mm. diameter, splitting into 5 (rarely 4) valves, the valves as many as the carpels, inflexed and remaining attached to the centre of the ovary. Seeds many, reniform, concentrically ridged. Fig. 11/1–2, p. 33.

KENYA. Northern Frontier Province : Marsabit, on Moyale road, July 1942, *Mrs. J. Bally in Bally* 1842 !
DISTR. **K1** ; tropical and North Africa from the Atlantic Islands and Cape Verde to Arabia and Baluchistan, in the drier regions ; also Portuguese East Africa, South and South West Africa
HAB. " In a small crater "

This species is very variable in the size of its parts and in its degree of hairiness.

12. DELOSPERMA

N.E. Br. in Gard. Chron., ser. 3, 78 : 412 & 433 (1925) in clave ; & in Burtt-Davy, Man. Fl. Pl. Transv. 1 : 157 (1926)

[*Mesembryanthemum* sensu auct., *non* L., Sp. Pl. : 480 (1753) sensu stricto]

Decumbent or prostrate succulent perennial herbs, usually branched with distinct internodes and often forming clumps, sometimes tuberous and stemless. Leaves opposite, sessile, lanceolate or linear-lanceolate, soft, green, minutely and distinctly papillose, devoid of radiating setae at their tips. Flowers solitary, terminal or axillary. Calyx gamosepalous, 5-lobed, the lobes unequal, the largest one or two horned or tailed and leaf-like. Staminodes many, obvious, petaloid, lanceolate, free. Stamens many, free, like the staminodes inserted on the calyx-tube. Ovary inferior, of 5 united carpels, 5-celled ; stigmas 5, free, style-like ; ovules and seeds many per cell, parietal. Fruit a 5-valved capsule, opening under moist conditions by means of 2 contiguous expanding keels running down the inner face of each valve, the keels membranous-winged in their apical (outer) part. Cell-wings absent, the cells thus unroofed and the seeds exposed in the expanded capsule.

About 130 species, about 5 in tropical Africa, the rest confined to South Africa.

Ovary 4-locular ; leaves ovate-cordate, flattened . . . *Aptenia cordifolia* (see note below)

Ovary 5-locular ; leaves sausage-shaped, subcylindrical or bluntly trigonous :

Papillae all (or all but very few) rounded or at most conical, so that young stems, pedicels and calyces appear glabrous :

Main branches 2·5–5 mm. in diameter, stout ; calyx-tube about 14 mm. long and 9 mm. in diameter across its mouth . . . . 1. *D. nakurense*

Main branches 2 mm. in diameter or less, weak ; calyx-tube 8–9 mm. long and 4–5 mm. in diameter across its mouth . . . 2. *D. abyssinicum*

Papillae of young stems, pedicels and calyces each with an apical hair-like process, giving these parts a pubescent appearance . . . 3. *D. oehleri*

NOTE. *Aptenia cordifolia* (Linn.f.) N.E. Br. has been included in the above key since it is commonly cultivated as a drought-resistant border plant and if found as an escape will come out to *Delosperma* in the key to genera. It can be recognized as a prostrate glabrous herb with ovate-cordate leaves, magenta or rosy-purple flowers, a four-locular inferior ovary, and a four-locular capsule without cell-wings or wings to the keel margins, and with axile placentation. It is a native of South Africa.

1. **D. nakurense** (*Engl.*) *Herre* in Sukkulentenk. 2 : 38 (1948). Type : Kenya, Nakuru, *Engler* 2023 (B, holo.)

A bushy decumbent somewhat woody succulent glabrous perennial herb ; branches stout, 10–60 cm. or more in length. Leaves sessile, subterete, papillose, fleshy, more or less patent, 12–35 mm. long, 1·5–3 mm. broad, somewhat dilated and connate in pairs at the base. Flowers solitary, on pedicels 0·5–1·5 cm. long, terminal and held above the level of the uppermost leaves, but often later overtopped by a lateral branch, white or pink, about 2·5 cm. across when expanded. Calyx about 14 mm. long, and 9 mm. in diameter across the top of the tube ; lobes triangular, about twice as long as broad, unequal, fleshy. Staminodes petaloid, lanceolate, white or pink, about 12 mm. long, 0·8 mm. broad. Fig. 10/1–3.

KENYA. Naivasha District : Longonot, Mar. 1922, *Dummer* 5162 ! ; Masai District : mile 26, [about 41.5 km.] Narok–Ngare Ngare, 17 June 1956, *Verdcourt* 1509 !
TANGANYIKA. Masai District : Olmerungu, Kimasai, 13 Jan. 1936, *Greenway* 4331 !
DISTR. **K**1, 3, 6 ; **T**2, 3 ; not known elsewhere
HAB. Amongst rocks in upland grassland, 1200–2200 m.

SYN. *Mesembryanthemum nakurense* Engl. in E.J. 43 : 195, fig. 6, M–R (1909)

Plants from Handeni and Lushoto Districts appear slightly different in habit, having shorter more numerous branches and rather longer-stalked flowers as compared with specimens from the central areas, but material is insufficient to warrant any taxonomic distinction of these forms. An example is :—Lushoto District, W. Usambara Mts., Lasa Mt., 6 Sept. 1953, *Greenway* 4049 !

2. **D. abyssinicum** (*Regel*) *Schwantes* in Gartenfl. 77 : 69 (1928). Type : cultivated in Leningrad from seeds sent by Schimper from Ethiopia (?LE, holo.)

A dwarf prostrate succulent herb. Branches subwoody-succulent, weak, green and leafy in younger parts, brown and leafless in the older, 20–40 cm. long or more. Leaves subcylindrical, sausage-shaped, sessile, connate at the base, fleshy, like the calyces and young stems crystalline-papillose, 10–23 mm. long, 2–4 mm. diameter. Flowers terminal, solitary, purple,

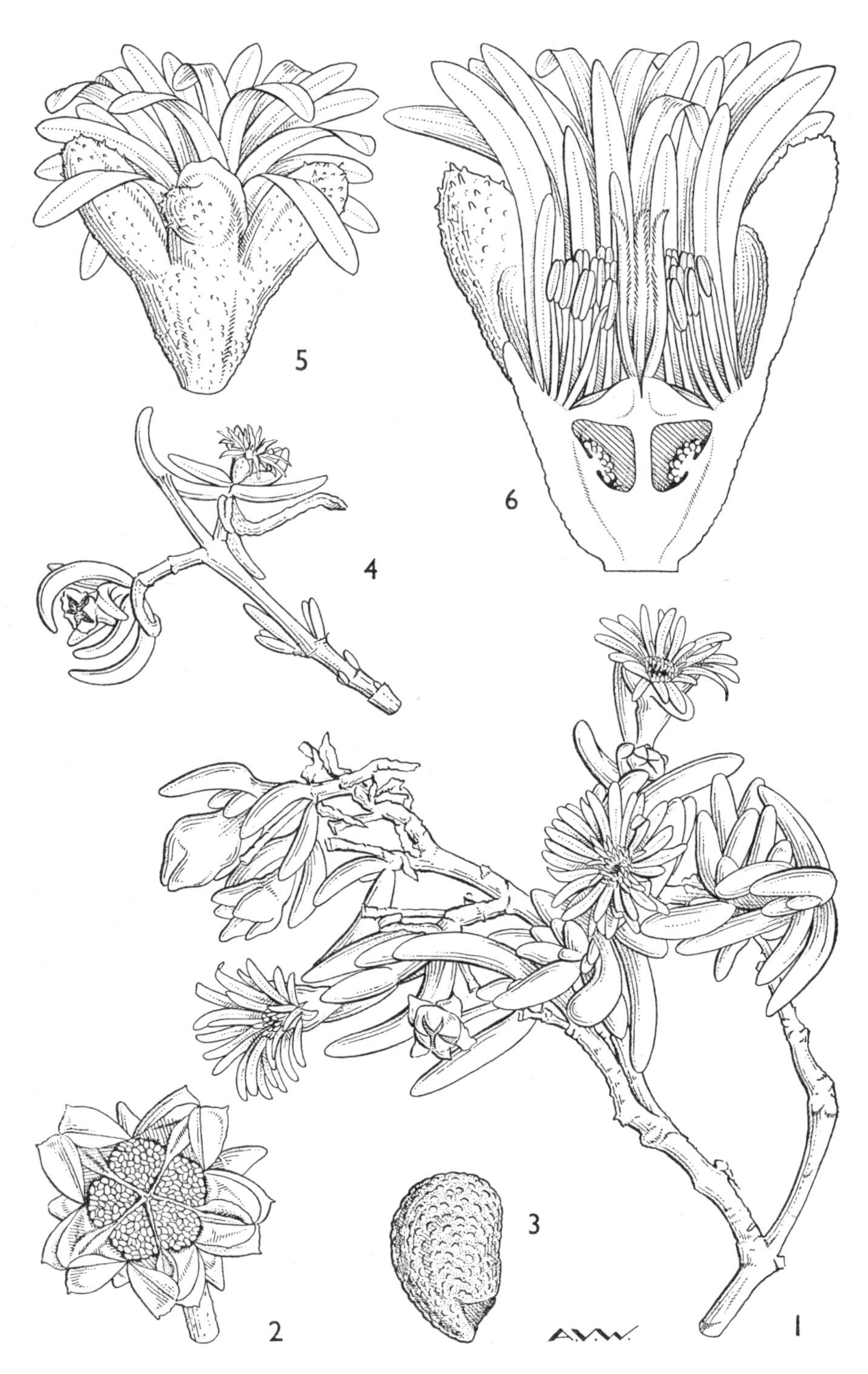

Fig. 10. *DELOSPERMA NAKURENSE*—**1,** habit, × 1 ; **2,** open capsule, × 2 ; **3,** seed, × 30 ; *D. ABYSSINICUM*—**4,** habit, × 1 ; **5,** flower, × 4 ; **6,** flower, in longitudinal median section, × 6. 1, from a photograph by *Bally ;* 2–3, from *Greenway* 4049 ; 4, from *Bally* 5643 ; 5–6, from *Gillett* 13948.

magenta or pink. Calyx 8–9 mm. long, 4–5 mm. in diameter across the top of the tube ; lobes fleshy, horned, unequal, one very small, two larger, and the largest about twice the length of the intermediate pair, and resembling the leaves. Fig. 10/4–6, p. 31.

KENYA. Northern Frontier Province : Mt. Kulal, 14 Oct. 1947, *Bally* 5643 ! ; Furroli, 20 Sept. 1952, *Gillett* 13948 (a pink-flowered form)
DISTR. **K1** ; Ethiopia and Eritrea
HAB. Rock ridges and crannies ; 2000–2100 m.

SYN. *Mesembryanthemum abyssinicum* Regel in Gartenfl. 22 : 299, t. 774 (1873)

D. nakurense may not be specifically distinct from this species, which is insufficiently well-known for a true assessment of its size and colour variation to be made.

3. **D. oehleri** (*Engl.*) *Herre* in Sukkulentenk. 2 : 38 (1948). Type : Tanganyika, Masai District, Ndassekera Mts., N. of Lamuniane, *Oehler & Jaeger* 365 (B, holo., K, iso. !)

A succulent cushion-forming herb. Branches procumbent, then ascending, subwoody-succulent. Leaves lanceolate, succulent, 14–20 mm. long, 3 mm. broad, 2·5 mm. thick, 3-angled, canaliculate above, basally truncate. Flower terminal, held above the level of the leaves, shortly-stalked. Calyx about 10 mm. long and 6 mm. in diameter across the top of the calyx-tube, the tube rounded below, pubescent-papillose. Calyx-lobes unequal, triangular, about 1½ times as long as broad. Staminodes lanceolate, gradually narrowed to the base, about 7 mm. long.

TANGANYIKA. Masai District : Ndassekera Mts., N. of Lamuniane, 8 Jan. 1907. *Oehler & Jaeger* 365 !
DISTR. **T2** ; known from this single gathering
HAB. Not known

SYN. *Mesembryanthemum oehleri* Engl. in E.J. 43 : 194, fig. 6, F–L (1909) ; F.D.O.-A, 2 : 260 (1938)

Efforts should be made to relocate and conserve this rare plant.

13. TETRAGONIA

L., Sp. Pl. : 480 (1753) & Gen. Pl., ed. 5 : 215 (1754) ; Adamson, J. S. Afr. Bot. 21 : 109 (1955)

Often minutely papillose annual or perennial herbs or subshrubs. Leaves alternate, entire, subsucculent. Stipules absent. Flowers axillary, solitary or fasciculate, sessile or pedunculate, hermaphrodite or unisexual, greenish or yellowish. Calyx gamosepalous, the tube produced above the ovary, 3–5-lobed. Stamens few to many, alternate with the calyx-lobes, solitary or fasciculate. Ovary semi-inferior or inferior, 1–9-locular ; ovules 1 per loculus, pendulous ; styles as many as the loculi, linear, free. Fruit simple, dry, hard, indehiscent, terete, winged, ridged, horned or spiny.

Probably some 60 species, of which about 40 are South African, the rest mostly Australasian and in western South America ; 2 or 3 species in Somaliland and Socotra.
There is much to be said for treating this and the following genus as members of a separate family, the *Tetragoniaceae*, more closely allied to the *Nyctaginaceae* and *Gyrostemonaceae* ; see Friedrich in Phyton 6 : 220–263 (1956).

Fruit spiny . . . . 1. *T. acanthocarpa*
Fruit not spiny . . . . 2. *T. tetragonoïdes*

FIG. 11. *AÏZOÖN CANARIENSE*—**1,** habit, × 1 ; **2,** fruit, in plan, × 4 ; *TETRAGONIA ACANTHOCARPA*—**3,** habit, × 1 ; **4,** fruit, × 4 ; *TRIBULOCARPUS DIMORPHANTHUS*—**5,** habit, ×1 ; **6,** inflorescence, × 3. 1–2, from *Mrs. J. Bally in Bally* 1842 ; 3–4, from *Bogdan* 3044 ; 5–6, from *Gillett* 13384.

1. **T. acanthocarpa** *Adamson* in J. S. Afr. Bot. 21 : 146 (1955). Type : South Africa, *Gill* 206 (BOL, holo.)

A semi-succulent papulose annual herb. Stems prostrate, 8–30 cm. long. Leaves petiolate ; blade elliptic or rhombic, obtuse at the apex, decurrent into the petiole at the base, 9–45 mm. long, 4–24 mm. broad. Petiole 1–20 mm. long. Flowers axillary, solitary, pedicellate. Calyx-segments 4, obtuse. Stamens 4. Ovary inferior, 3-(sometimes 2- or 4-) locular. Fruit 3–7 mm. diameter, densely but rather softly spiny. Fig. 11/3–4, p. 33.

KENYA. Nakuru District : Lake Elmenteita, 16 June 1951, *Bogdan* 3044!
DISTR. **K**3, 4 ; South Africa ; introduced in Kenya
HAB. Upland grassland ; 1500–1600 m.

2. **T. tetragonoïdes** (*Pallas*) *O. Ktze.*, Rev. Gen. 2 : 264 (1891). Type : cultivated in Moscow, from seeds sent by Jacquin (location of type uncertain).

A succulent erect or trailing annual herb. Stems bright green, up to 100 cm. long or more. Leaves ovate-rhombic, obtuse or bluntly acuminate at apex, cuneate and more or less decurrent into the petiole at the base, petiolate ; blade 15–110 mm. long, 10–75 mm. wide ; petiole 5–25 mm. long. Flowers solitary or paired, subsessile, axillary. Calyx-segments 3–5, usually 4, green outside, yellow-green inside, unequal. Stamens 4–22, solitary or fasciculate. Ovary semi-inferior, of 5–8 united carpels. Fruit smooth, 5–10 mm. long, bearing a horn below each calyx-segment, the horn sometimes giving rise to another flower or a branchlet.

UGANDA. West Nile District : Arua Station, Apr. 1940, *Eggeling* 3900!
KENYA. Naivasha District : Gilgil, *Thorold*!
DISTR. **U**1 ; **K**3–4 ; native of New Zealand—" New Zealand Spinach " ; widely introduced by cultivation as a vegetable.
HAB. Abandoned cultivations, where it may persist for a time.

SYN. *Demidovia tetragonoïdes* Pallas, Hort. Demid.: 150, t. 1 (1781)
Tetragonia expansa Murr., Comm. Götting 6 : 13, t. 5. (1783), *nom. illegit.* Type: as *T. tetragonoïdes* (Pallas) O. Ktze.

14. TRIBULOCARPUS

S. Moore in J.B. 59 : 228 (1921)

Farinose shrublets. Leaves alternate, exstipulate. Flowers in pedunculate leaf-opposed capitula, congested, polygamous, the upper one hermaphrodite, the lateral ones hermaphrodite or male. Calyx gamosepalous, the tube very short in male flowers, 3–5 mm. long and cylindrical in hermaphrodite flowers ; segments 5. Stamens many, inserted in the mouth of the calyx-tube. Ovary inferior, syncarpous, of 2 carpels, 2-locular ; ovules 1 per cell, pendulous ; style 1, stigmas 2. Fruit compound, spiny.

Monotypic.

T. dimorphanthus (*Pax*) *S. Moore* in J.B. 59 : 288 (1921) ; Verdc. in K.B. 12 : 348 (1957). Type : South West Africa, *Marloth* 1249 (B, holo., PRE, iso.)

A grey-green densely papillose subshrub up to 50 cm. high. Leaves obovate-spathulate or obovate, entire, retuse, rounded or obtuse, often mucronulate, at apex, cuneate at the base, petiolate. Blade 4–25 mm. long, 4–14 mm. broad ; petiole 3–11 mm. long. Flowers sessile, white or yellow, congested, on a peduncle 3–30 mm. long. Fruit subspherical, spiny, the spines representing indurated bracts. Fig. 11/5–6, p. 33.

KENYA. Northern Frontier Province : 20 km. SSW. of El Wak, 29 May 1952, *Gillett* 13384 !

DISTR. **K1** ; adjoining areas of Ethiopia (Ogaden) and Somaliland ; also South West Africa and Namaqualand

HAB. In *Acacia-Commiphora* bushland ; about 400 m.

SYN. *Tetragonia dimorphantha* Pax in E.J. 10 : 12 (1888)

Plants from our area and its adjoining regions seem to be generally rather smaller in all parts, especially in fruit-size, than those from the type area, and perhaps may represent a distinct subspecies, but there is no other apparent difference.

INDEX TO AÏZOACEAE